THE COMPUTATION OF
CHEMICAL EQUILIBRIA

T0296185

THE
COMPUTATION
OF CHEMICAL
EQUILIBRIA

by

F. VAN ZEGGEREN, D.Sc.

and

S. H. STOREY, Ph.D.

CAMBRIDGE
AT THE UNIVERSITY PRESS
1970

CAMBRIDGE UNIVERSITY PRESS
Cambridge, New York, Melbourne, Madrid, Cape Town, Singapore,
São Paulo, Delhi, Dubai, Tokyo, Mexico City

Cambridge University Press
The Edinburgh Building, Cambridge CB2 8RU, UK

Published in the United States of America by Cambridge University Press, New York

www.cambridge.org
Information on this title: www.cambridge.org/9780521172257

First published 1970
First paperback edition 2010

A catalogue record for this publication is available from the British Library

Library of Congress Catalogue Card Number: 78–92255

ISBN 978-0-521-07630-2 Hardback
ISBN 978-0-521-17225-7 Paperback

CONTENTS

FOREWORD

This book represents a major contribution to the literature of the thermodynamics of complex chemical equilibria. It should be welcomed by workers already in the field because it provides a critical survey of the analytical techniques developed in many different countries and laboratories over the last quarter century. For those unfamiliar with these developments this book provides an excellent introduction to methods of great generality and power, which should find ever increasing application in the next few years. There is every reason to hope that the successes already achieved in high temperature reactions, explosive design and chemical processing can be extended and enlarged, even to such complex fields as biology and geology.

From the academic standpoint this book should also be well received. Thus for some years the graduate thermodynamics course for chemical and mechanical engineers at Washington University has included a discussion of complex chemical equilibrium problems. The presentation has been handicapped by the fact that the methods described in this book were not generally available in textbook form. While many fine texts, such as Denbigh's *Principles of Chemical Equilibrium* do discuss the problem of free energy minimization, to the student it always appears that the method, though general in principle, is confined in practice to systems in which the stoichiometric equations can be written and where each derived constituent is represented in the final set of equations by a chemical equilibrium or mass action equation. While this method is certainly valid the problem is more instructively considered and conceptually easier to approach by disregarding the details of the processes. The total free energy of the system can be considered in terms of the chemical potentials of the reactants and a set of possible products and finding the composition at the minimum of the multidimensional free energy-composition surface. The method is limited only by the size of the computational facilities available and by the number of chemical compounds and temperature ranges for which standard chemical potentials are tabulated. It is useful for the student to learn that thermodynamic data are now tabulated (albeit in sometimes obscure locations) for many compounds even to temperatures in the 20,000 °K range. Discussion of the origin of the numbers in these tables is in itself a most instructive pastime but for the student it becomes a matter of considerable interest to compute equilibrium compositions in such exotic reactions as those between diborane and oxygen-bifluoride, at say, 6,000 °K. The ease with which temperature, pressure and starting compositions can be varied can almost be said to add a new dimension to the instruction of students in the thermodynamics of chemical equilibrium.

Although this book could thus serve as a supplementary text for advanced thermodynamics courses it is primarily designed for those actively working in the field. If a complaint could be made about this book it is a complaint that can be levelled against much of the published literature—there are too few papers in which the theoretical calculations are compared with experimental results. Perhaps this book will inspire the publication of observations which can be checked against the associated computations. This becomes particularly important when questions of chemical kinetics may be significant, as for example in shock tube studies. This raises speculations beyond the scope of this work but which may well be of major concern in certain particular cases; the equilibrium compositions will, however, always be needed. In any event, there is reason to believe that this book in its own way will help to make the next twenty-five years of study of complex reactions as productive and interesting as the last.

E. B. BAGLEY

St Louis, Missouri
November 1968

PREFACE

It is difficult to overestimate the effect of the increased availability of electronic digital computers on the field of chemical equilibrium computation over the last fifteen to twenty years. The last and only comprehensive monograph in this field was the publication by Kobe and Leland, in 1954. Since that time, computers have assumed more and more the often prodigious numerical tedium involved in equilibrium calculations. Also, with the increasing importance of the problem in high temperature, high pressure processing, in rocketry, and in explosives technology, a revival of interest in the equilibrium problem has led to the accumulation of a considerable body of additional technical literature. The work of S. R. Brinkley and his collaborators, in the late 1940s, laid the groundwork for systematic treatments of the problem. These have replaced the older semi-intuitive manual methods. The range of applications is amply demonstrated by the number of different journals in which descriptions of novel techniques have appeared. In fact, owing to the lack of a recent comprehensive treatment of the various techniques now available, and to the wide spread in the literature, some duplication has already occurred.

It is hoped that this book will be of use in three areas. First, since there is now a sufficient variety of methods available to allow one to be chosen to fit a given problem, the book provides a guide to the methods available and their properties, for those who have specific problems to solve. Secondly, the fundamental chemical thermodynamic and numerical analytic material has been selected so as to make the book suitable as a graduate level text, in particular for students in Chemical Engineering faculties. Finally, this book should represent, for a short while, a summary of the current position for workers in the field. It is with all three aims in mind that the notation has been made as consistent as possible throughout the book, and an attempt has been made to indicate under which problem conditions some of the published methods are less unsatisfactory than others.

The authors would like to express their thanks to Canadian
Industries Limited and to the University of Liverpool, for permis-
sion to write the book, and for allowing the use of computational,
library and office facilities during its preparation. Particular thanks
are due to Professor A. Young and to his successor, Professor
M. R. Sampford, of the Department of Computational and
Statistical Science of the University of Liverpool, for their en-
couragement and for their interest in this work. The authors are
much indebted to Professor E. B. Bagley, of the Department of
Chemical Engineering of Washington University, St Louis, Mo.,
who read the book in draft and made many useful comments, and
for writing a foreword to this book. Many others have been
connected with the work at various times; numerous computations
have been carried out over the past six years by Mrs M. L. Taylor,
Mrs N. Brownlie and Mrs A. D. Lovie, Mr H. R. Hughes has
given valuable clerical assistance. Mrs F. Mang, Mrs J. F.
O'Connor and Miss N. Prout have performed excellently the
very considerable task of converting the authors' handwriting into
the final typescript. To all of these the authors would like to extend
their grateful thanks.

F. v. Z.
S. H. S.

1

THE FOUNDATIONS OF CHEMICAL
EQUILIBRIUM COMPUTATION

1.1 Prologue

1.1.1 The history of equilibrium computation

Until the early 1940s, only two methods appear to have been in use for the computation of chemical equilibria.

The first of these was based on the assumption that only a few molecular species would be present in the final equilibrium composition. By making use of mass action relationships, a limited number of non-linear equations specifying the composition were obtained. These were then solved numerically with such techniques as were available in the standard numerical analysis texts of the time (e.g. Scarborough, 1930).

The second method employed a cumbersome and time-consuming trial and error approach, in which the user's intuition played a crucial part. In fact, the solution of the problem often depended entirely on the user's intuition.

Furthermore, all such calculations were done on the calculators that were available until that time and those were, by present-day standards, very unsophisticated and mainly mechanical in nature. The principles first enunciated by Babbage in the nineteenth century had not yet resulted in the development of electronic, decision-making digital computers.

The development of rockets in the Second World War, particularly in Germany, led to a reconsideration of computational procedures for the solving of chemical equilibrium problems, because the technology of rockets such as the V2 required an accurate knowledge of the chemical equilibria of the propulsion gases formed in the chemical decomposition of the propellants within the rockets. Several articles describe the attempts at obtaining more accurate and, in particular, faster computational procedures for finding chemical equilibria. It is not surprising to find that two authors working at the Hermann Goering Air Force Research Institute in Germany published, as early as 1943, a paper entitled: 'The Composition of Dissociating Gases and Calculation of Simultaneous Gas Equilibrium'. The article was published as a Goering Institute Report, but also as a more generally available article (Damköhler and Edse, 1943). These authors describe a method wherein the non-linear equations, obtained on insertion of mass action equations into mass balance equations, are solved by a graphical estimation technique. They apply their method to an equilibrium mixture of gases, consisting of the elements carbon, hydrogen, nitrogen and oxygen, at specified pressure and temperature. For this system the procedure consists in estimating two starting values, viz. the partial pressures of H_2O and of O_2, or, instead of the latter, the partial pressure ratio of CO_2 to CO can often be used advantageously. The other partial pressures are then calculated from these two initial estimates with the appropriate equilibrium constants. Changes in initial estimates are made until both the desired total pressure and the total oxygen balance equations are satisfied.

Algebraic procedures, very similar in nature to the graphical one mentioned in the previous paragraph, were proposed after computers became available (Winternitz, 1949; Donegan and Farber, 1956; Harker, 1967). Several later articles describe methods wherein all composition variables are related to one estimation variable (Von Stein and Voetter, 1953; McEwan, 1950).

The above mentioned schemes suffer from a number of serious disadvantages. The main problem is that the procedure employs mass action and mass balance equations written specifically for the system under consideration, and these equations have to be re-written for each new system. Furthermore, the procedure does not always succeed in converging to the desired equilibrium composition. Also, the amount of time involved in graph plotting, particularly when many different sets of pressure and temperature have to be investigated, is usually prohibitive.

The first major advance was made after the Second World War and most of the credit for the development of a generalized scheme is due to S. R. Brinkley. In 1946, Brinkley paved the way for a systematic approach to equilibrium computation, by giving an analytical criterion for the number of independent components in a multi-constituent system (see §1.3.2). Subsequently, the same author published the outline of a method with which equilibria could be computed (Brinkley, 1947). The method is described in considerable detail in §4.2, so it should suffice to state here that Brinkley's procedure is basically a method wherein, first, a number of components is selected (see §1.3.2), with which, through the use of chemical equilibrium constants, the concentrations of all other chemical species in the equilibrium mixture are computed. These 'derived' species concentrations then serve as corrections to the atom balance equations and the correction procedure consists in solving a set of these corrected equations, with as many unknowns as there are elements (or components). Brinkley applied several methods, amongst others the standard Newton–Raphson linearization method for solving such sets of equations. Brinkley's method requires less set-up time than the method described by Damköhler and Edse, and furthermore, the finding of the numerical solution of a given problem, for instance that of the combustion of propane, actually requires much less time than the graphical method described above. But, and this is the main contribution that Brinkley's method made, his calculation scheme is very amenable to the use of digital computers. It is perhaps not entirely a coincidence that Brinkley's method first achieved success at the time when digital computers made their entry into the market and became generally available to research institutions, particularly in the United States.

Wilkes (1956) gives an excellent summary of the history of digital computers. He describes in some detail how Babbage came to stop the development of his analytical engine in 1843, and how the first model of the 'Automatic Sequence Controlled Calculator' was conceived by H. H. Aiken in 1937 and completed, at Harvard University, in 1944. This device was operating mechanically, but only two years later, in the summer of 1946, the first electronic computer (ENIAC) was built at the Moore School of Electrical Engineering at Pennsylvania State University. The same school built the EDVAC, wherein the von Neumann principle, of using words for both storage and commands, was utilized. This principle gives modern computers their real power in that they can modify their own behaviour to cope with changes in external or internal conditions. The first computer to be designed (at Cambridge University) and used in England was the EDSAC, in May 1949. From the late 1940s until the present time, in a time-span of only two decades, the advances in computer technology have been almost unbelievably rapid. The extreme speed with which calculations could be performed led to the development of a whole new body of science: computer science, of which the field of equilibrium computation may be considered to form part.

During the last twenty years, fast and efficient methods for calculating equilibrium compositions have led to developments in three major areas of applied science. These are (see §1.1.2):

1. The development of rocket propellants.
2. The scientific evaluation of explosives.
3. The development of sophisticated chemical processing techniques, particularly high pressure and high temperature processes.

Of the three, the advance in propellant technology in particular has led to the development of a large number of sometimes very ingenious techniques for the solution of simultaneous chemical equilibrium problems. It is even possible that the development of propellants and rocket and space science, by creating a demand for rapid computation, contributed considerably to the development of computers sketched in the above paragraph. Certainly, only the development of larger and faster computers has made possible the development of propellants and rockets as they are in use today. Both these developments have had their main impetus in the United States, and is for this reason that most of the references to computational procedures in this monograph are to papers published in the United States (or in Canada).

1.1.2 Applications of equilibrium calculations

As was pointed out in §1.1.1, there has been a large interaction between computational procedures on the one hand, and applications of such computations on the other. There have been, during the past two decades, four main areas of applications.

(a) The main application, and one that is important from the point of view of illustrating the need for accurate chemical equilibria, is the calculation of properties of propellants and rocket motors. One of the most important characteristics of propellants in a rocket motor is the specific impulse, which determines the apogee of the rocket that can be attained. The specific impulse is directly related to the exit velocity of the gas from the nozzle attached to the rocket motor. The exhaust velocity is determined by the difference ΔH in enthalpy H between nozzle exit and chamber:

$$I_s = \frac{V_e}{g} = \frac{\sqrt{(\Delta H)}}{g}, \qquad (1.1.1)$$

where g is the gravitational constant. Thus, the enthalpies of the gas mixtures at the nozzle exit and in the chamber have to be calculated and since these depend on the chemical compositions of the respective gas mixtures, it can be seen that it is necessary to evaluate the equilibrium compositions of these mixtures. Solving these problems is now a standard exercise for propulsion engineers (Sutton, 1963). Computation of chemical equilibria is done by a standard subroutine in computer programs for the computation of specific impulse and of many other significant rocket design parameters and characteristics. Such important characteristics are, e.g. thrust, mass flow rate (propellant consumption), nozzle design, etc.

(b) Another application is found in the calculation of properties of explosives. The standard thermohydrodynamic theory of explosives (Cook, 1958) requires the knowledge of the thermodynamic properties of the gas mixture at the Chapman–Jouguet plane at the end of the reaction zone accompanying the detonation (or reactive shock) wave. These thermodynamic properties can only be found after having calculated the chemical equilibrium of the gas mixture in the reaction zone. In explosive property calculations, the problem is more complex than in the calculation of propellant or rocket properties and characteristics, because explosives give gas mixtures at extremely high pressures (of the order of millions of p.s.i.) and temperatures (several thousand °K). The equation of state describing such high pressure and temperature mixtures is one that cannot yet be predicted on the basis of theoretical considerations, such as the virial equation of state. The virial equation of state has been applied, but only for standard mixtures consisting of the elements carbon, hydrogen, nitrogen and oxygen; for most other mixtures empirical equations of state have to be used (see §1.4.1). The non-ideality of the explosion products makes it necessary to develop special techniques for equilibrium computation (van Zeggeren and Storey, 1969).

(c) There are numerous examples of applications of complex equilibrium calculations in chemical processing. With the rapidly advancing technology that chemical processing has experienced over the past ten years, particularly with the increase in the number of high pressure, high temperature processes with their attendant rapid chemical kinetics, and in view of the increasing complexity and size of chemical processing plants, a large impetus has been given to rapid, computer control of such plants. Computer control usually requires a knowledge of the exact compositions encountered at each stage of the process under consideration. Also, with increase in competition, optimization of processes to give either maximum yield, efficiency, safety or any other extreme property, has led to the development of a considerable volume of optimization theory (Wilde and Beightler, 1967). In the optimization of chemical processes, chemical equilibria must be computed. It is rather interesting to note that, conversely, several techniques developed as optimization methods can be used as methods for calculating equilibria (see §3.2). Typical examples of high temperature, high speed commercial processes are combustion processes, such as in power plants, magnetohydrodynamic developments, plasma jet techniques, etc.

(d) The fourth main area of application is to problems of such complexity that manual calculation could never be attempted, even with all possible approximations and simplifications. Typical of such applications are the RAND methods for calculating the behaviour of multiphase biological cell systems (see e.g. Dantzig and DeHaven, 1962). Some applications have even been made in studies on the geological origin of organic materials (Eck et al. 1966); the study in question indicated that organic compounds such as aromatics may be formed, under conditions of thermodynamic equilibrium, from inorganic ingredients such as carbonaceous chondrites, at moderate temperatures and low pressure. It seems that the next logical step in such studies could well be the investigation of the origin of life. (For an interesting and stimulating discussion of chemical evolution, see Calvin, 1965). It is hoped that some of the methods described in this book will be able to cope with problems of such formidable complexity.

1.2 Thermodynamics of chemical equilibria

1.2.1 Introduction

The computation of chemical equilibria is essentially a problem which relies on the solving of equations which can be derived from classical thermodynamics. In this section an outline will be given of those thermodynamic principles which are usually encountered in the solving of the computation problem. Although the thermodynamic principles are equally valid for ideal gases, non-ideal gases, and condensed compounds, and for mixtures thereof, this section will deal only with ideal gases and gas mixtures. The next section (1.3) will treat heterogeneous systems, and §1.4 will illustrate some of the added complexities encountered in the treatment of non-ideal gaseous systems.

1.2.2 Dalton's Law

The equation of state for ideal gases can be written as (Zemansky, 1957):

$$pV = nRT, \tag{1.2.1}$$

where n is the number of moles of all gases present in the system, thus:

$$n = n_1 + n_2 \ldots + n_N = \sum_{i(g)=1}^{N} n_i, \tag{1.2.2}$$

where N is the number of chemical species present in the (gaseous) system.* Equation (1.2.1) can also be written as:

$$p = \sum_i \frac{n_i RT}{V}, \tag{1.2.3}$$

$$p = p_1 + p_2 + \ldots p_N = \sum_i p_i, \tag{1.2.4}$$

---

* The word species here denotes a distinct chemical constituent in a given phase. The same constituent in another phase will be considered as the same species.

where it is postulated that p_i can be called a partial pressure of the system, thus: $p_i = (n_i RT/V)$. Equation (1.2.4) is known as Dalton's law (Zemansky, 1957). Equations (1.2.1) and (1.2.4) can be combined into

$$p = \sum_i \frac{n_i}{n} p = \sum_i p_i. \qquad (1.2.5)$$

A useful value is the mole fraction of the ith gas which is denoted by x_i, viz. $x_i = n_i/n$, etc. Also $p_1 = x_1 p$, etc. It follows that

$$\sum_i \frac{n_i}{n} = x_1 + x_2 \ldots x_N = 1 \qquad (1.2.6)$$

so that, of the total number of N mole fractions, only $N-1$ are independent of each other; the last one can be determined from the other $N-1$ fractions.

1.2.3 Entropy and Gibbs free energy of gas mixtures

The entropy of an ideal gas (species i) at temperature T and pressure p is given by (Zemansky, 1957):

$$(S_T)_i = (S_{298})_i + \int_{298}^{T} (C_p)_i \frac{dT}{T} - R \ln p. \qquad (1.2.7)$$

For a number of different gases i, prior to mixing, the total entropy at p and T is equal to the sum of each of the entropies of the gases:

$$S_T^n = \sum n_i (S_{298})_i + \int_{298}^{T} \sum_i n_i (C_p)_i \frac{dT}{T} - \sum n_i R \ln p, \qquad (1.2.8)$$

where the superscript u indicates: unmixed. After mixing, assuming no chemical reaction takes place, the total entropy of the mixture is, according to Gibbs, equal to the sum of the partial entropies. The partial entropy s_i is the entropy that gas i would have if it occupied the whole volume alone at T, in which case it would exert a pressure equal to p_i. Thus:

$$S_T^m = \sum_i n_i (S_{298})_i + \int_{298}^{T} \sum_i n_i (C_p)_i \frac{dT}{T} - \sum n_i R \ln p_i, \qquad (1.2.9)$$

where the superscript m indicates: mixed. Thus, from equations (1.2.8) and (1.2.9) follows an expression for the entropy of mixing:

$$(\Delta S)_{\text{mixing}} = -\sum n_i R \ln \frac{p_i}{p} \qquad (1.2.10)$$

which, according to equation (1.2.5) becomes:

$$(\Delta S)_{\text{mixing}} = - \Sigma\, n_i R \ln x_i \qquad (1.2.11)$$

which has always a positive value: the spontaneous mixing process is accompanied by an entropy increase. This is in accordance with the second law of thermodynamics (Zemansky, 1957).

The Gibbs free energy G (also called the Gibbs function or the free enthalpy) is given by:

$$G_T = H_T - TS_T. \qquad (1.2.12)$$

Thus, since $H_T = H_{298} + \displaystyle\int_{298}^{T} C_p\, dT$, equation (1.2.12) becomes:

$$G_T = H_{298} - TS_{298} + \int_{298}^{T} C_p\, dT - T\int_{298}^{T} \frac{C_p}{T}\, dT + RT \ln p. \qquad (1.2.13)$$

Zemansky (1957) writes this expression as follows:

$$G_T = RT\,(\psi + \ln p), \qquad (1.2.14)$$

where ψ is a function of T only. In a manner similar to that described for the entropy, the Gibbs free energy of mixing can be derived. One obtains:

$$(\Delta G)_{\text{mixing}} = RT \sum_i n_i \ln x_i. \qquad (1.2.15)$$

In this case, $(\Delta G)_{\text{mixing}}$ is always negative, in accordance with the knowledge that mixing is a spontaneous process.

Many other thermodynamic variables such as the volume, enthalpy and specific heats, are purely additive for ideal gas mixtures (Prigogine and Defay, 1954). Thus, for these variables:

$$(\Delta V)_{\text{mixing}} = 0, \quad (\Delta H)_{\text{mixing}} = 0 \quad \text{and} \quad (\Delta C_p)_{\text{mixing}} = 0.$$

Equations (1.2.14) and (1.2.15) can be combined to give the frequently used expression for G at p and T for any (ideal) gas mixture:

$$G = RT \sum_i n_i\,(\psi_i + \ln p + \ln x_i). \qquad (1.2.16)$$

1.2.4 Chemical equilibrium conditions

A chemical equilibrium can be defined as the thermodynamic state of a system which, for a given set of thermodynamic variables, is unalterable in its mechanical, thermal and chemical properties. The conditions for chemical equilibrium can be derived as follows: Consider a system of constant mass, in mechanical and thermal but not chemical equilibrium. The system is in contact with its surroundings at temperature T, and undergoes an infinitesimal irreversible process involving an exchange of heat dQ with the surroundings. If dS_{sys} and dS_{sur} are the changes in entropy of system and surroundings, respectively, the total entropy change of system + surroundings is $dS_{sys} + dS_{sur}$ and since the process is irreversible this sum has to be greater than zero. Thus:

$$dS_{sys} + dS_{sur} > 0.$$

The entropy change of the surroundings is due merely to entropy transfer (Lewis and Randall, 1961), thus:

$$dS_{sur} = \frac{dQ_{sur}}{T} = -\frac{dQ_{sys}}{T}.$$

Then, by deleting the subscript 'sys' one obtains for the system:

$$dQ - T\,dS < 0. \qquad (1.2.17)$$

In the irreversible process, the internal energy of the system changes by dE, and the amount of work performed is $p\,dV$. Thus, according to the first law:

$$dE + p\,dV - T\,dS < 0. \qquad (1.2.18)$$

This is a very fundamental inequality in thermodynamics, that leads to the basic condition for chemical equilibrium, viz. that $dE + p\,dV - T\,dS = 0$ at equilibrium. Usually, the irreversible process is restricted in that two of the thermodynamic variables are maintained constant. The usual selection of these two variables is pressure and temperature. For p and T constant, inequality (1.2.18) becomes (Zemansky, 1957)

$$d(E + pV - TS) < 0 \quad \text{or} \quad dG < 0, \qquad (1.2.19)$$

or, in other words, the Gibbs free energy of a system at constant p and T decreases during an irreversible process, becoming a minimum at the final equilibrium state. Another selection of variables, viz. that of V and T being kept constant, leads to

$$d(E - TS) < 0 \quad \text{or} \quad dA < 0. \qquad (1.2.20)$$

A is the free energy (or Helmholtz free energy, or work function) which becomes a minimum at chemical equilibrium for the case where V and T are specified.

1.2.5 Chemical potentials

In subsequent sections of this chapter all derivations will be based on Gibbs free energy, G, but it should be remembered that for many systems a consideration of these derivations for A is often more useful and equally valid.

Infinitesimal changes in chemical equilibrium lead to corresponding changes in G or A. G is a function of all independent variables of a system. Thus: $G = G(p, T, n_i)$ and

$$dG = \left(\frac{\partial G}{\partial T}\right)_{p,n_i} dT + \left(\frac{\partial G}{\partial P}\right)_{T,n_i} dp + \sum_i \left(\frac{\partial G}{\partial n_i}\right)_{p,T,n_l} dn_i.$$

$$(1.2.21)$$

It can be shown (Zemansky, 1957) that:

$$\left(\frac{\partial G}{\partial T}\right)_{p,n_i} = -S \quad \text{and} \quad \left(\frac{\partial G}{\partial p}\right)_{T,n_i} = V.$$

So that $\qquad dG = -SdT + VdP + \sum_i \left(\frac{\partial G}{\partial n_i}\right)_{p,T,n_l} dn_i.$ $\qquad$ (1.2.22)

The partial differentials $(\partial G/\partial n_i)_{p,T,n_l}$ are intensive properties of the system, indicating the effect on the Gibbs free energy of the system when a small amount of species i is added to the system. The derivative is usually called the chemical potential (μ_i) of the ith species; it is a function of p, T, *and all other* n_i's. Thus, equation (1.2.21) can be written in abbreviated form:

$$dG = -SdT + Vdp + \sum_i \mu_i dn_i.$$

$$(1.2.23)$$

It is useful to relate chemical potentials to standard chemical potentials (at a partial pressure of 1 atm.), in a manner similar to that for other thermodynamic functions. For example, the expression (1.2.23) for dG, at constant temperature and n_i becomes, on integration, from 1 atm. to p_i:

$$G_T - G_T^\circ = \int_1^{p_i} V dp,$$

$$(1.2.24)$$

where the superscript $^\circ$ denotes the standard condition (1 atm.).
Then, using the ideal gas equation of state in the form:

$$V = \frac{n_i RT}{p_i} \qquad (1.2.3; \ 1.2.5)$$

leads to: $\qquad G_T - G_T^\circ = n_i \, RT \ln p_i \qquad (1.2.25)$

and subsequent differentiation with respect to any n_i gives:

$$\mu_i = \mu_i^\circ + RT \ln p_i. \qquad (1.2.26)$$

Thus, this expression now gives physical significance to the
function ψ_i in equation (1.2.16), since at given p and T:

$$G = \sum_i \mu_i \, n_i \qquad (1.2.27)$$

and with (1.2.26) this leads to:

$$G = \sum_i n_i \, (\mu_i^\circ + RT \ln p_i) \qquad (1.2.28)$$

from which follows that $\psi_i = \mu_i^\circ / RT$.

For additional properties of chemical potentials, and their
relation to component potentials in reactive systems, see §2.3.

1.2.6 Equations of reaction equilibrium

Consider a system consisting of various numbers of moles of
different chemical species, in thermal and mechanical but not
chemical equilibrium. We will not be concerned here with the rate
of reaction, but we will assume that the chemical species react with
each other and, in so doing, proceed towards a chemical equilibrium. We will also assume that the reaction proceeds under
constant pressure and constant temperature conditions so that, as
above, we consider the Gibbs free energy change of the system,
and the system moves towards a situation where the Gibbs free
energy is a minimum.

Any general reaction between many chemical species can be
considered as a series of simpler reactions, as discussed in chapter 4.
A typical simple reaction would be represented by the equation:

$$\sum_i \nu_i \mathscr{A}_i \rightarrow \sum_l \nu_l \mathscr{A}_l, \qquad (1.2.29)$$

where ν_i and ν_l are reaction coefficients, which can be both positive and negative numbers of moles, and $\mathscr{A}_i$ and $\mathscr{A}_l$ represent the molecular formulae of the species i and l respectively. Reaction (1.2.29) will proceed to an extent ϵ corresponding to the final chemical equilibrium state. For any ϵ, there are present in the system

$$\sum_i \nu_i(1-\epsilon) \quad \text{and} \quad \sum_l \nu_l \epsilon \qquad (1.2.30)$$

numbers of moles of species $\mathscr{A}_i$ and $\mathscr{A}_l$ respectively. Then, from equation (1.2.27), the Gibbs free energy of the system becomes:

$$G = \sum_i \nu_i(1-\epsilon)\,\mu_i + \sum_l \nu_l\,\epsilon\mu_l. \qquad (1.2.31)$$

It has been shown that equilibrium corresponds with a minimum value of G, so that $\partial G/\partial\epsilon = 0$, and from equation (1.2.31):

$$\sum_i \nu_i\,\mu_i = \sum_l \nu_l\,\mu_l. \qquad (1.2.32)$$

Substituting the expression (1.2.26) into equation (1.2.32), one obtains:
$$\sum_i \nu_i(\mu_i^\circ + RT \ln p_i) = \sum_l \nu_l(\mu_l^\circ + RT \ln p_l)$$

or
$$\ln \frac{\prod_l (p_l)\,\nu_l}{\prod_i (p_i)\,\nu_i} = \frac{-(\sum_l \nu_l\mu_l^\circ - \sum_i \nu_i\mu_i^\circ)}{RT}. \qquad (1.2.33)$$

The left-hand side is usually denoted by $\ln K_p$, where K_p is the equilibrium constant for reaction (1.2.29) expressed in terms of partial pressures. The right-hand side of the equation shows that the equilibrium constant for a given chemical reaction is a function only of temperature, since the μ_i° and μ_l° values are standard state values, dependent on temperature only. In fact, the difference in the numerator of the right-hand side of equation (1.2.33), i.e.

$$\sum_l \nu_l\mu_l^\circ - \sum_i \nu_i\,\mu_i^\circ$$

is the standard Gibbs free energy change of reaction (1.2.29), denoted by $(\Delta G^\circ)_r$. Therefore, we can now write equation (1.2.33) as follows:
$$-\frac{(\Delta G^\circ)r}{RT} = \ln (K_p)_r, \qquad (1.2.34)$$

where the subscript r refers to the reaction under consideration. Equation (1.2.34) is a mass action equation, of great importance in the evaluation of chemical equilibria by one of the two main approaches in chemical equilibrium computations. Standard free energy data, as a function of temperature, are available for most chemical species, usually as standard Gibbs free energy of formation values, for the formation of the chemical species from the assigned reference elements (see §1.5).

1.3   Heterogeneous systems

1.3.1   Introduction

Lewis and Randall (1961) define a homogeneous system as a system whose properties are the same in all parts, or in other words: in which there are no apparent surfaces of discontinuity. Thus, usually, a system consisting of one or more gases, and which is at thermal and mechanical equilibrium is a homogeneous gas system. A heterogeneous system is a system consisting of two or more homogeneous regions each one of which is called a phase. The phases are separated from each other by surfaces of discontinuity.

For a heterogeneous system, consisting of Φ phases, and of N chemical species distributed over these phases, the Gibbs free energy can be written as (Zemansky, 1957):

$$G = \left(\sum_{i=1}^{N} \mu_i n_i \right)_{\text{phase 1}} + \left(\sum_{i=1}^{N} \mu_i n_i \right)_{\text{phase 2}} + \ldots \quad (1.3.1)$$

For Φ phases, and utilizing ϕ as a superscript, equation (1.3.1) becomes:

$$G = \sum_{\phi=1}^{\Phi} \left(\sum_{i=1}^{N} \mu_i^{\phi} n_i^{\phi} \right). \quad (1.3.2)$$

Similarly, by expressing the dG^{ϕ} according to equation (1.2.23) for each homogeneous phase, and summing dG^{ϕ} expressions, one obtains (using the same limits on i and ϕ as in (1.3.2) throughout):

$$dG = -S\,dT + V\,dp + \sum_{\phi} \left(\sum_i \mu_i^{\phi}\,dn_i^{\phi} \right), \quad (1.3.3)$$

where S and V are the entropy and volume of the whole system. As before, the condition for chemical equilibrium is that G be a minimum, so that $dG = 0$, and at constant T and p, equation (1.3.3) becomes:

$$\sum_{\phi} \left(\sum_i \mu_i^{\phi}\,dn_i^{\phi} \right) = 0. \quad (1.3.4)$$

The dn_i^ϕ values are not all independent, but are related to each other via constraint equations. For the general case of multiphase, multiconstituent systems, without or with chemical reactions and phase transfers of the constituent species, a study of these constraint equations leads to the famous phase rule, originally formulated in 1875 by Gibbs. However, Gibbs' description of the derivation of this rule (Gibbs, 1961) is so short as to be almost meaningless for the purpose of its application to the problems of chemical equilibrium computation. A very elegant and rigorous derivation of the phase rule (due to Brinkley, 1946) can be given. Some of the aspects of this derivation are described in §1.3.2 below. Additional details are given in chapter 2. In §1.3.3 a discussion is given of the utilization of the phase rule, and particularly of the associated concept of components, in chemical equilibrium computation.

1.3.2 Derivation of the Phase Rule

Consider first a system of N species in Φ phases, wherein all species are considered to be *inert*. All phases are in equilibrium with each other, in other words: transport of any species from one phase to another can freely take place. Thus, in the expression for the Gibbs free energy G (1.3.2) the n_i^ϕ are subject to the constraints:

$$\sum n_i^\phi = n_i = \text{constant} \quad (i = 1, \ldots, N), \qquad (1.3.5)$$

which express that the mass of species i is conserved, regardless of its distribution over the different phases.

The condition of equilibrium then becomes that of minimizing G subject to the constraints (1.3.5). The minimization is most elegantly handled by the method of Lagrangian multipliers. (For a simple example of the use of this technique, see Appendix I.) The Lagrangian function for the problem is:

$$L = \sum_\phi \sum_i \mu_i^\phi n_i^\phi + \sum_i \chi_i \left\{ n_i - \sum_\phi n_i^\phi \right\}, \qquad (1.3.6)$$

where the χ_i are Lagrangian multipliers. Minimization of L then leads to the equations:

$$\mu_i^1 = \ldots \mu_i^\phi = \ldots = \mu_i^\Phi = \mu_i = \chi_i \quad (i = 1, \ldots, N).$$
$$(1.3.7)$$

Equations (1.3.7) express the physical condition that the chemical potential of a species is to be the same in all phases, if equilibrium is to exist in the whole system. Thus, equations (1.3.7) are the equations of phase equilibrium; there are $N(\Phi-1)$ equations of this type. In addition, for each phase there is a relationship $\sum_i x_i^\phi = 1$ (equation 1.2.6); hence there are Φ such expressions. All these expressions connect the thermodynamic variables x_i^ϕ and, for instance, p and T; thus there are $N\Phi+2$ variables. The difference between the number of variables and the number of relations between these variables, is called the variance, or number of degrees of freedom, (F):

$$F = N-\Phi+2. \qquad (1.3.8)$$

This is the phase rule of Gibbs, but, as mentioned above, in the form of equation (1.3.8) it only applies to systems of inert species.

For systems in which some or all of the constituents are chemically active the general treatment becomes more complex. It is beyond the scope of this chapter to give a detailed derivation of the general phase rule, although in the description of the mathematical structure underlying the chemical equilibrium problem, the equations that are derived result directly in the phase rule (§2.3). However, it is fruitful to consider, at this stage, and in as rational and systematic a manner as possible, the stoichiometry of a chemical system. Brinkley (1946) has presented such a systematic approach. A general, and, as far as notation is concerned, somewhat simpler description of a chemical system was given by Aris (1965). Aris uses the summation convention from tensor analysis which is adopted in the following.

Consider a system of N species, in chemical equilibrium. A molecular species or compound $\mathscr{A}_i$, is an entity of the form:

$$\mathscr{A}_i = a_i^e \mathscr{B}_e \quad (i = 1, ..., N), \qquad (1.3.9)$$

where the a_i^e are integers, denoting the number of atoms of element $\mathscr{B}_e$ in molecule $\mathscr{A}_i$. The $\mathscr{B}_e$ are symbols for the elements or atomic species, $(e = 1, ..., M)$. The elements themselves are molecular species with $a_i^e = \delta_i^e$, the Kronecker delta. For every i, the a_i^e define a formula vector α_i. If the rank of the matrix of vector elements a_i^e is C, it follows that there are C linearly independent vectors (denoted by c), and $N - C$ linearly dependent vectors which may be expressed as linear combinations of the a_c^e:

$$v_i^c a_c^e = a_e^i. \qquad (1.3.10)$$

These correspond to the chemical reactions:

$$v_i^c \mathscr{A}_c = \mathscr{A}_i. \qquad (1.3.11)$$

Thus, the specification of C species is sufficient for the description of the system. These C species are called *components* of the system, and since generally $C < N$, the set of components is a subset of the set of species.

The concept of components is particularly useful in expressing the phase rule for a chemically reactive, multiconstituent, multiphase system. The phase rule for such a system can be derived (see §2.3), viz:

$$F = C - \Phi + 2. \qquad (1.3.12)$$

The rank of the matrix of the a_i^e for a chemical system at equilibrium is usually equal to the number (M) of chemical elements $\mathscr{B}_e$ in that system. However, there are two exceptions to this rule. One of these exceptions is a trivial one, which finds its origin in the fact that a system may contain species, whose formula vectors are such that any two elements appear only in the same ratio in the compounds of which they form part (e.g. in polymerization and isomerization). The number of such constraints may be called Z. The other exception occurs when the system to be considered is only a pseudo-equilibrium, i.e. a system at equilibrium with some restrictions imposed by the kinetics of the reactions necessary to reach equilibrium. This problem has been considered by several authors in some detail (Prigogine and Defay, 1947; Peneloux, 1949; Schott, 1964), and its solution is of considerable importance in studies on comparisons between computed and experimentally found equilibrium compositions. The number of independent reactions subject to such constraints may be called R'. Then, the number of components is given by:

$$C = M - Z + R'. \qquad (1.3.13)$$

This equation is equivalent to the expression derived by Zemansky (1957):

$$C = N - R - Z, \qquad (1.3.14)$$

where R is the number of independent reactions that *can* take place between the various constituents, and thus:

$$N - M = R + R'. \qquad (1.3.15)$$

Equation (1.3.12) expresses the phase rule in the form in which it is to be used in subsequent chapters, and it will be assumed that $C = M$, unless it is stated explicitly that C is given by (1.3.13).

1.3.3   Utilization of the Phase Rule

The phase rule expression (1.3.12) is of very great importance, and mainly because of this rule, thermodynamics has had a considerable impact on the other physical sciences, ever since 1875, when the phase rule was first formulated by Gibbs.

The variance F has been given a very precise meaning. For a non-variant system ($F = 0$), p and T are fixed, and the composition, in all phases, is determined as well. For a mono-variant system ($F = 1$), there is one variable too many, and the equilibrium of the system is not determined until one of the variables is arbitrarily chosen.

In the derivation of the phase rule (§1.3.2) it was tacitly assumed that every constituent was present in every phase. However, this restriction can be removed quite simply, because for every species i not present in a given phase ϕ, the corresponding equilibrium equation $\mu_i^\phi = \chi_i$ is lacking. Thus, both the number of variables and the number of equations have been reduced by one and F remains unchanged (Brinkley, 1946; Zemansky, 1957). This important conclusion leads to the possibility of considering a multi-constituent gas phase in equilibrium with a number of pure condensed phases. The above reasoning shows that, for such systems, the phase rule applies equally well as it does to the more general case.

As was shown in §1.3.2 above, the number of components, C, is usually equal to the number of elements M, and this equality will be assumed to apply throughout the rest of the book. Also, in many chemical equilibrium problems, the condensed phases are assumed to be pure compounds and not mixtures. Assuming that in all problems a gas phase is present, and denoting the number of condensed species by S, the number of phases is given by:

$$\Phi = S + 1. \qquad (1.3.16)$$

Then, the phase rule becomes:

$$F = M - S + 1. \qquad (1.3.17)$$

The number of degrees of freedom, F, is equal to the number of thermodynamic variables to be specified (usually two, see §1.6.2), plus the number of ratios of elemental abundances (§2.2), n_R, that can be arbitrarily specified:

$$F = 2 + n_R. \qquad (1.3.18)$$

Combining equations (1.3.17) and (1.3.18) leads to:

$$S = M - 1 - n_R \qquad (1.3.19)$$

and this indicates that, for each condensed species present in the system, one fewer ratio of elemental abundances needs to be specified. Equation (1.3.19) leads to the important conclusion that the maximum number of condensed species, in the particular systems considered here, is given by

$$S_{\max} = M - 1. \qquad (1.3.20)$$

When this is the case: $n_R = 0$, or in other words: one atom balance equation suffices for the computation of the complete chemical equilibrium of the gas phase. There will always be at least one atom balance equation, or a linear combination of atom balance equations, with which such a computation can be carried out, since $M - S \geqslant 1$.

Equation (1.3.20) is of considerable use in computations of equilibria in systems where the number of condensed species may be expected to be close to M, e.g. in extractive metallurgy (Kubaschewski and Evans, 1958). If one uses a method such as, e.g. Naphtali's direct gradient technique (see §3.4.1) to calculate the equilibria in such multi-solid/gas phase systems, the estimates with which the computation proceeds are rather critical, because negative concentrations of solids can be predicted at equilibrium or convergence can be slow. The rule expressed in equation (1.3.20) allows the procedure to select as few as possible (S_{max}) of all possible solids, and by permutation and combination to find the best possible set of starting estimates quite quickly.

In the derivation of the phase rule, in §1.3.2, several of the derived auxiliary equations are of considerable importance in that they provide a physical basis for some of the apparently purely mathematical techniques described in chapters 3 and 4.

The equations of phase equilibrium (1.3.7) are, for a system of one species in two phases:

$$\mu_1^{(1)} = \mu_1^{(2)}.$$

For this system, at constant p and T, equation (1.3.3) becomes:

$$dG = \mu_1^{(1)} dn_1^{(1)} + \mu_1^{(2)} dn_1^{(2)}$$

and since $dn_1^{(2)} = -dn_1^{(1)}$, this leads to:

$$dG = (\mu_1^{(1)} - \mu_1^{(2)}) dn_1^{(1)}.$$

The condition is that, at equilibrium, G become a minimum. Thus, in going from a non-equilibrium state to an equilibrium state, dG must be negative. This means that a mass flow of the species 1 from phase (1) to phase (2) is accompanied by a decrease in $\mu_1^{(2)}$. Conversely (Zemansky 1957): If the chemical potential in one phase is larger than that in another phase, equilibrium will be reached by mass flow from the one phase to the other. In this sense, the chemical potential is a thermodynamic variable equivalent to temperature gradient (which generates a flow of heat to reach equilibrium) and to pressure gradient (which generates a 'flow' of work to reach equilibrium). This concept can be used to advantage in iteration procedures (see §§2.6 and 3.4.2).

Another important conclusion can be drawn from the expression that all μ_i^ϕ are equal to the Lagrangian multiplier χ_i corresponding to the mass constraint for species i (equation 1.3.7). This is that the Lagrangian multipliers, which are auxiliary mathematical aids, are also of physical significance. This significance is discussed further in subsequent chapters. The expression $\mu_i = \chi_i$ explains the basic equivalence of the two groups of methods described in chapters 3 and 4.

1.4 Non-ideal gases and gas mixtures

1.4.1 Equation of state

An ideal or perfect gas is a gas which fulfills two conditions (Lewis and Randall, 1961). One condition is that its energy is a function of temperature only or, in other words, that its dependence on volume (or pressure) is zero:

$$\left(\frac{\partial E}{\partial V}\right)_T = 0. \qquad (1.4.1)$$

The second condition is that when the temperature, pressure and volume of the gas are changed, these variables obey the relationship

$$pV = nRT, \qquad (1.4.2)$$

where n is the total number of moles of the ideal gas. Equation (1.4.2) is the equation of state for an ideal gas. For any substance an equation of state exists relating pressure, volume and temperature. The equation of state is an experimental relationship which is observed to be obeyed for such substances. The ideal gas law (equation 1.4.2) is a limiting law which applies only to hypothetical substances (actual substances are never really ideal). However, the applicability of the ideal equation of state depends on the accuracy required of any thermodynamic properties calculated with the aid of that equation. At high temperatures and/or pressures in the gas state, the accuracy given by equation (1.4.2) is usually not sufficient. Many other, more accurate, experimental relationships have been expressed in analytical form similar to equation (1.4.2). In general the equation of state may be written in the form

$$p(V - \alpha) = nRT, \qquad (1.4.3)$$

where α, the covolume, is a function of volume and temperature, or pressure and temperature, or pressure and volume. These are all really equivalent and in the general case we may write $\alpha = \alpha(V, T)$. Such properties as fugacity, enthalpy, free energy, entropy, specific heat, etc., are related to integrals or derivatives of the equation of state functions. Therefore, it is desirable to express α as accurately as possible in a mathematical form which can be integrated or differentiated readily to yield the other thermodynamic functions.

The number of equations of state, expressible in the form
(1.4.3), which have been published in the literature are extremely
numerous. It is not within the scope of this volume to discuss all
or even a considerable number of these equations of state. We
shall only deal with three equations of state which are often
encountered and which can be readily incorporated in most of
the methods for the computation of equilibrium compositions
described in this volume. The best known of these is the virial
equation of state, wherein a power series is used to express the
covolume of a gas, thus:

$$pV = A + \frac{B^1}{V} + \frac{C^1}{V^2} + \frac{D^1}{V^3} + \dots, \qquad (1.4.4)$$

where A, B^1, C^1, D^1, ... are the first, second, third, fourth, etc.
virial coefficients, respectively. The virial coefficients are functions
of temperature, and express the effects of the mutual interactions
between the different molecular or atomic species within the gas
system. Equation (1.4.4) was first proposed by Kamerlingh Onnes
in 1901, and has found considerable use since then in expressing,
analytically, relationships between high pressure, low temperature
thermodynamic state variables. A large number of experimental
data on virial coefficients is available in the literature. Also, there
is now considerable theoretical justification for expressing the
deviations from ideality in these virial coefficient forms. Equation
(1.4.4) can also be written in the form:

$$pV = nRT + Bp + Cp^2 + Dp^3 + \dots \qquad (1.4.5)$$

and it has been customary, for gaseous systems at moderate
pressures, to truncate this equation as follows:

$$pV = nRT + Bp, \qquad (1.4.6)$$

where $B = B(T, x)$ can be expressed in terms of the sum of the
interactions between the various types of species:

$$B(T, x) = \sum_i \sum_l B_{il}(T) x_i x_l, \qquad (1.4.7)$$

where now $B_{il}(T)$ is a function of temperature only, for the
interaction between species i and l.

Another equation of state is one wherein α is expressed as a function of volume only, thus

$$\alpha = \alpha(V). \tag{1.4.8}$$

This form of the equation of state has found application in explosives technology where it was assumed by M. A. Cook (1958) that there is no interaction between the different species within the explosion product compositions. The equation, as used by Cook, has one considerable advantage in that it also shows no dependence of energy on volume, at constant temperature; thus, equation (1.4.1) applies here as well. Also, there is no effect of pressure on specific heat at constant volume, so that the thermochemical calculations utilizing this equation of state become very simple indeed. The α, V function has been determined for a large variety of different explosive product compositions by Cook and it has been found, within experimental error, that a single equation is obtainable for such explosive mixtures. Equation (1.4.8) cannot, without prior investigation, be applied to other than explosive product compositions.

A third equation of state, applicable to gases at moderate pressures, i.e. up to 1,000 atm, that expresses the relationship between pressure, volume and temperature with sufficient accuracy for most applications is the van der Waals equation of state:

$$\left(p + \frac{a}{V^2}\right)(V - b) = nRT, \tag{1.4.9}$$

where a and b are constants, independent of temperature. For a pure gas, a and b are characteristic constants. For a mixture of gases, a and b can be written in terms of the a_i and b_i values of the various species in the mixture, viz. (Prigogine and Defay, 1954):

$$a = (\sum_i a_i^{\frac{1}{2}} x_i)^2 \tag{1.4.10}$$

and

$$b = \sum_i b_i x_i. \tag{1.4.11}$$

Equation (1.4.10) represents the interaction of forces between dissimilar molecules; equation (1.4.11) represents the assumption that the volumes of the molecules themselves are purely additive. Prigogine and Defay discuss, in detail, the ranges of applicability of the van der Waals equation, and show that for pressures up to about 1,000 atm. the equation gives remarkably satisfactory accuracy.

1.4.2 Fugacities

The term fugacity was first introduced by Lewis, in 1901 (Lewis and Randall, 1961). Lewis and Randall give a historical and formal description of the derivation and significance of fugacity for gases, liquids and solids, but for our purposes it is more convenient to use a different means of introduction for the concept of fugacity. We have already seen (1.2.26) that for an ideal gas:

$$\mu = \mu^\circ(T) + RT \ln p. \tag{1.2.26}$$

Prigogine and Defay call a system ideal when it obeys the equation (1.2.26). They then write, for any non-ideal gas, the analogous equation:

$$\mu = \mu^\circ(T) + RT \ln f, \tag{1.4.12}$$

where $f = f(T, p)$ is the fugacity of that gas. At zero pressure all gases are perfect, and f becomes equal to p. At any finite pressure, a relationship between f and p can be derived, viz. (Prigogine and Defay, 1954):

$$\left(\frac{\partial \ln f}{\partial p} \right)_T = \frac{V}{nRT}. \tag{1.4.13}$$

The actual expression for f as a function of T and p can be obtained by integration of equation (1.4.13), after insertion of the appropriate equation of state on the right-hand side. If, e.g. the truncated virial equation (1.4.6) is used, equation (1.4.13) leads to:

$$\ln \frac{f}{p} = \frac{Bp}{RT}. \tag{1.4.14}$$

For the Cook and van der Waals equations of state, the expressions for $\ln(f/p)$ are more complicated and will not be given here. However, numerical solutions are easily obtainable by stepwise integration.

For mixtures of non-ideal gases the treatment of fugacity is completely analogous. Prigogine and Defay use the defining equation:

$$\mu_i = \mu_i^\circ + RT \ln f_i \tag{1.4.15}$$

for a species i in the mixture. Here, f_i is the partial fugacity. As for a pure, non-ideal gas, one can derive an expression relating f_i to p, viz:

$$\left(\frac{\partial \ln f_i}{\partial p}\right)_T = \frac{V_i}{nRT}, \tag{1.4.16}$$

where V_i/n is the partial molar volume of species i. The determination of partial fugacities therefore depends on a knowledge of partial volumes as a function of pressure. Experimental data are scarce, particularly for multi-component non-ideal gas mixtures. For simplicity, use is often made of Lewis' rule, which states (Prigogine and Defay, 1954):

$$f_i = x_i f_i^\circ, \tag{1.4.17}$$

where f_i° is the fugacity of the pure species i at the temperature T and total pressure p at which f_i is to be evaluated. Equation (1.4.17) allows for an evaluation of f_i from f, which is analogous to the evaluation of p_i from p (see §1.2.2).

Some typical examples of applications of non-ideal equations of state to computations of chemical equilibrium compositions have been published, viz. by Boynton (1963) and by Michels and Schneiderman (1963). A description of their methods will be given in the appropriate sections in chapter 4. It will be seen that one of the main complicating factors in using the virial equation of state is to find an analytically expressible function of the interaction of the various first-order coefficients B_{il} between various molecules and atoms in the equilibrium composition. The virial treatment, even in its truncated form, is extremely complex and it is not surprising therefore that few treatments of non-ideality have been presented in the literature.

1.4.3 Thermodynamic excess functions

Equation (1.4.15) can also be written in the form:

$$\mu_i = \mu_i^\circ + RT \ln (p_i \gamma_i), \tag{1.4.18}$$

where the γ_i are the activity coefficients. In gaseous systems, the γ_i are also sometimes called fugacity coefficients. Equation (1.4.18) can be used for calculating equilibria in real mixtures, including liquid and solid solutions. In the case of condensed solutions, the activity coefficients are directly related to the osmotic coefficients, and these can be measured by a number of different experimental techniques (Prigogine and Defay, 1954).

The availability of activity coefficients makes possible the calculation of the effect of non-ideality on thermodynamic functions. In §1.2.3, expressions were given for the entropy and Gibbs free energy of mixing, for the case of ideal gas mixtures. Equation (1.4.18) provides an easy means of extending these thermodynamic functions of mixing to non-ideal systems. Instead of equation (1.2.15), one obtains for the change in Gibbs free energy on mixing:

$$(\Delta G)_{\text{mixing}} = RT \sum_i n_i \ln (x_i \gamma_i). \qquad (1.4.19)$$

The difference between the thermodynamic function of mixing for a real system, and the value corresponding to an ideal system at the same temperature and pressure, is called the thermodynamic excess function, denoted by the superscript E (Prigogine and Defay, 1954). Thus:

$$G^E = (\Delta G)^{\text{real}}_{\text{mixing}} - (\Delta G)^{\text{ideal}}_{\text{mixing}} = RT \sum_i n_i \ln \gamma_i, \qquad (1.4.20)$$

where G^E is the excess Gibbs free energy. Similar expressions are given by Prigogine and Defay for H^E, S^E, V^E and C_p^E, viz.

$$H^E = -RT^2 \left(\sum_i n_i \frac{\partial \ln \gamma_i}{\partial T} \right), \qquad (1.4.21)$$

$$S^E = -R \sum_i \left(n_i T \frac{\partial \ln \gamma_i}{\partial T} + n_i \ln \gamma_i \right), \qquad (1.4.22)$$

$$V^E = RT \sum_i n_i \frac{\partial \ln \gamma_i}{\partial p} \qquad (1.4.23)$$

and

$$C_p^E = -RT \sum_i \left(2 n_i \frac{\partial \ln \gamma_i}{\partial T} + T n_i \frac{\partial^2 \ln \gamma_i}{\partial T^2} \right). \qquad (1.4.24)$$

1.5 Thermodynamic data

As is shown in §1.6.2, the evaluation of chemical equilibrium compositions requires a knowledge of values of the standard enthalpy and standard entropy of the constituents in the system, both as a function of temperature. These thermodynamic functions are given, for each constituent, by the expressions:

$$H_T^\circ - H_{298}^\circ = \int_{298}^{T} C_p^\circ \, dT \qquad (1.5.1)$$

and
$$S_T^\circ - S_{298}^\circ = \int_{298}^{T} \frac{C_p^\circ}{T} \, dT, \qquad (1.5.2)$$

where the reference temperature is taken to be 25 °C = 298·15 °K, and where the superscript ° denotes standard state (i.e. 1 atm. pressure). Thus, an evaluation of H_T° and S_T° at any temperature T, requires a knowledge of the functions $C_p^\circ(T)$.

During the past two decades considerable advances have been made in the measurement and tabulation of thermodynamic data. The most comprehensive set of data can be found in the JANAF Thermochemical Tables (Stull, 1965). The JANAF (Joint Army Navy Air Force) tables have been compiled over a ten year period in a project sponsored by the Advanced Research Projects Agency. The Dow Chemical Company compiled the data, with D. R. Stull as editor, and a panel of scientists from government and industry collaborated to review the data critically prior to their publication and distribution. This approach was designed to ensure that the table be of the highest possible quality. The tables are given in loose-leaf form and are updated annually. The JANAF tables give values of the standard thermodynamic functions C_p°, S_T°, $-(G_T^\circ - H_{298}^\circ)/T$, $H_T^\circ - H_{298}^\circ$, ΔH_f°, ΔG_f°, and log K_{pf}, at 100 °K intervals, between 0 °K and 6000 °K, for several hundred species. Unfortunately, many of the tables give interim data, of doubtful accuracy, whereas all data are given to the same number of significant figures, regardless of their estimated accuracy. Thus, in order to establish their reliability and accuracy, all these data have to be weighed by the user prior to their application in the calculation of chemical equilibria. In a later chapter (see §2.6 and 5.4), the effect of errors in the thermodynamic data on the chemical equilibrium compositions, and the significance of such deviations, are discussed in some detail.

Several other compendia of thermodynamic data, at different temperatures, have been published recently, viz. by Landolt-

Börnstein (1961), Kelley (1960), and by NASA (McBride *et al.* 1963). A number of useful references are available dealing with specific groups of compounds, e.g. in the field of metallurgy (Kubaschewski and Evans, 1958), for fuels and oxidizers in propellants (Battelle Memorial Institute, 1949), for simple inorganic compounds (Bockris, White and Mackenzie, 1959), and for carbonates and sulphur compounds (Kelley, 1962). The last publication also gives heats of fusion and vaporization of a large number of inorganic substances. Also, standard thermodynamic values at the reference temperature, $298 \cdot 15°K$, have been published in recent years, e.g. enthalpy and entropy data by the U.S. National Bureau of Standards (Rossini *et al.* 1952; and partially updated: Wagman *et al.* 1968), entropy data by the U.S. Bureau of Mines (Kelley, 1961), and enthalpy data by Skinner (1964).

For many species, thermodynamic data over the range of interest are not available, and therefore it would be necessary to either disregard these species in the chemical equilibrium computations, or better, to obtain approximate estimated data. Several procedures are available in the literature for estimating thermodynamic properties of species. An interesting technique that is frequently used is the method of structural similarity, based on incremental group contributions. This method is particularly useful in calculations for large organic molecules (Janz, 1958). Also, for many simple compounds, estimation based on statistical thermodynamic calculations is possible; special computer programs which allow the user to do this, are now available, e.g. without charge from NASA (McBride and Gordon, 1967); a commercially available programme predicts a very comprehensive range of thermodynamic and physical data (American Institute of Chemical Engineers, 1965).

Modern computers have considerable memory storage space. However, it would be rather wasteful of such memory space if it were attempted to represent all thermodynamic data, such as those given in the JANAF tables, in the computer memory. This would require several hundred memory locations for each chemical species whose presence would be considered in the equilibrium calculations. Therefore, attempts must be and have been made to represent the various thermodynamic functions in the form of equations, whose coefficients can be stored much more compactly. Equations (1.5.1) and (1.5.2) show that one of the simplest ways to store data is to represent $C_p^\circ(T)$ as a polynomial function of T, for each species. Then, storage of H_{298}° (or rather of $H_{298}^\circ - H_0^\circ$), S_{298}°, and of the polynomial coefficients would suffice for the calculation of all thermodynamic data of that species, at any temperature T. The fitting of $C_p^\circ(T)$ to polynomial functions seems to be almost a tradition, although it would be just as logical, useful and accurate to fit either $H_T^\circ - H_{298}^\circ$, or $S_T^\circ - S_{298}^\circ$, and to obtain $C_p^\circ(T)$ from these fitted polynomials by differentiation. In every case, a conventional least-squares fitting technique can be applied, and standard subroutines are now available, for use on almost any computer, to allow a user to do this. However, Zeleznik and Gordon (1961) have pointed out that fitting only one function does not reproduce the thermodynamic data of the other two functions so that the residuals of all three functions are a minimum in the least-squares sense. These authors describe a method wherein a better, true least-squares, simultaneous fit to all three functions is obtained.

Wiederkehr (1962) shows how thermodynamic properties (H_T°, S_T°, G_T°) of multi-constituent systems can be represented conveniently in matrix form, wherein one matrix contains the coefficients of a fifth-order polynomial describing the $(C_p^\circ)_i$ function of temperature, for all species in the system. A similar matrix representation is given by Desré et al. (1964), who include condensed species, and their phase transitions as well.

For equilibrium computations at specified p (or V) and T, the only requirement is for Gibbs free energy or equilibrium constant data (see §1.6.2), and storage of Gibbs free energy data is easier than storage of either enthalpy or entropy data, because use can be made of the fact that the Gibbs free energy function:

$$-\left[\frac{G_T^\circ - H_{298}^\circ}{T}\right] \qquad (1.5.3)$$

is a smoothly varying function in the variable T, and interpolation of tabulated data, at 500 °K intervals, gives fairly accurate values for the Gibbs free energy at any temperature T. One disadvantage of this method is that the tabulated data are made up of values of S_{298}°, $H_{298}^{\circ} - H_{0}^{\circ}$, and the C_{p}°, and any change occasioned by the periodic updating of such values renders the tabulations useless. An elegant solution to this problem is given by Alcock (1966), who defines a Gibbs free energy function deviation $(GDF)_{T}$ as follows:

$$(GDF)_{T} = \frac{1}{T}(H_{T}^{\circ} - H_{298}^{\circ}) - (S_{T}^{\circ} - S_{298}^{\circ}) \qquad (1.5.4)$$

or, taking again 298·15 °K as a reference point,

$$(GDF)_{T} = \frac{1}{T}\int_{298}^{T} C_{p}^{\circ}\,dT - \int_{298}^{T} C_{0}^{p}\frac{dT}{T}. \qquad (1.5.5)$$

If C_{p}° is represented by the expression (Kelley, 1960):

$$C_{p}^{\circ} = a + bT + c/T^{2} \qquad (1.5.6)$$

the expression for the Gibbs free energy function deviation becomes:

$$(GDF)_{T} = a(1 - \theta + \ln\theta) + bT\left(\theta - \frac{1}{2} - \frac{\theta^{2}}{2}\right) + \frac{c}{T^{2}}\left(\frac{1}{\theta} - \frac{1}{2\theta^{2}} - \frac{1}{2}\right),$$

$$(1.5.7)$$

where $\theta = 298/T$. The functions in θ are universal functions to all species, whereas the constants a, b, and c differ for different species. Thus, a tabulation of these θ functions, at 500 °K intervals, will allow an accurate estimation of the corrections $(GDF)_{T}$ to be applied to ΔG for any reaction from:

$$\Delta G_{T}^{\circ} = \Delta H_{T}^{\circ} - T\Delta S_{298}^{\circ} + T\Delta(GDF)_{T}. \qquad (1.5.8)$$

The only data to be stored for the species are therefore the values of a, b, c, ΔH_{298}° and ΔS_{298}°, and, as Alcock points out, these data can easily be changed whenever new experimental data become available.

A modification in the previously described method by Alcock was introduced by Shelton and Blairs (1966). In this modification, allowance is made for changes of state of the species in the reaction considered, between 298 °K and T °K. The additional data required for the species are then the enthalpy and temperature (or entropy) of transformation.

It was shown above (see equation 1.2.33) that the value of the equilibrium constant for a given reaction can be obtained from values of the μ_i° and μ_l° of the species i and l participating in that reaction. The standard chemical potential values are also required in calculating equilibria via the free energy minimization methods to be described in chapter 3. There is some confusion in the literature on equilibrium computation as to which values to use for the μ_i°. For instance, White *et al.* (1958), and Oliver *et al.* (1962) use the expression:

$$\mu_i^\circ = (G_T^\circ - H_{298}^\circ)_i + (\Delta H_{f,298}^\circ)_i. \qquad (1.5.9)$$

However, this does not agree with the convention (Glasstone and Lewis, 1960) which states that the Gibbs free energy of all elements in their standard states is arbitrarily taken as zero at all temperatures. Thus: G_T° of any compound is equal to ΔG_f° at any T, and for the elements $G^\circ = \Delta G_f^\circ = 0$. Since for a pure compound $\mu_i^\circ = G^\circ$, we obtain:

$$\mu_{T,i}^\circ = (\Delta G_{f,T}^\circ)_i \qquad (1.5.10)$$

and for all reference elements:

$$(\mu_T^\circ)_i = 0. \qquad (1.5.11)$$

Equations (1.5.10) and (1.5.11) are much simpler to use than equation (1.5.9), particularly since most references (such as the JANAF tables) give values of ΔG_f° of species i at equal intervals of T. The expression (1.5.9) is of course equally valid, because the actual values of μ_i° are unimportant; only their differences are thermodynamically defined, and the selection of μ_i° values is based on convention only. In the examples discussed in chapter 5, all μ_i° values are defined by equations (1.5.10) and (1.5.11).

1.6 Computation of chemical equilibria

1.6.1 Existence and uniqueness of solutions

It will be assumed throughout the remainder of the book that each chemical equilibrium problem does, indeed, have a solution, and that such a solution is unique. Thus, every equilibrium problem can be solved, and the solution of the problem will be the only one corresponding with the thermodynamic variables specified for the particular problem under study (§1.6.2).

Several proofs of existence and uniqueness theorems have been given in the literature on chemical equilibria. A simple theorem, due to Aris (1965), shows that a solution is unique if the Jacobian, made up of coefficients ($\partial \ln (K_p) r / \partial \epsilon_r$), does not vanish (cf. §1.2.6). Hancock and Motzkin (1960) have, in rather more detail, developed the necessary and sufficient conditions for both existence and uniqueness of a solution to a standard problem, viz. that of finding the chemical equilibrium of a gas-pure solids system. Subsequently, Shapiro and Shapley (1964) have given proofs of existence and uniqueness theorems for the solutions of (a) equations arising from mass action laws, and (b) a minimization procedure (cf. §2.1). Shapiro and Shapley have made a rather thorough analysis of the precise relationship between the two approaches. In developing their theorems, the authors invoke a number of geometrical arguments. These arguments will not be considered here. However, it is interesting to note that very similar arguments can be used in developing convergence theorems, in particular for minimization procedures. This is done, for instance, in an application to equilibrium computation of an iterative procedure originally developed for solving convex programming problems (Warga, 1963).

In a recent article (Shear, 1968) it is shown that there cannot be a continuum of equilibrium points for a chemical reaction system. Shear invokes kinetic arguments to prove this; he then shows that there cannot be a second equilibrium point anywhere in the reaction simplex, because that would imply a continuum of equilibrium points. Shear also uses Brouwer's fixed-point theorem in proving that there must be at least one equilibrium point reached in any kinetic, i.e. chemical reaction, system. Thus, each reaction system has one, and only one, equilibrium point.

1.6.2 Thermodynamic variables

According to the phase rule, as expressed by equation (1.3.12), the number of degrees of freedom of a reactive chemical system is:

$$F = C - \Phi + 2. \tag{1.6.1}$$

The number of elemental abundances that can be independently specified (as ratios), is equal to $C-1$ for a gaseous system (see equation 1.3.18) and equal to $C-\Phi$ for a general, heterogeneous system. Thus, the remaining independent variables that can and must be specified for a complete determination of the chemical system is equal to $(C-\Phi+2)-(C-\Phi) = 2$. These two independent variables are the thermodynamic variables, to be selected from the set: p, V, T, E, H, S, A, G.

Three groups of specifications are most frequently met in chemical equilibrium problems:

1. p (or V) and T: This group represents calculations for systems for which the equilibrium is to be calculated regardless of the chemical system from which the equilibrium system is to be derived. By calculation of the equilibrium composition at p (or V) and T it is possible to calculate all other thermodynamic functions. In the minimization procedure (chapter 3), either Gibbs free energy G or free energy A is used.

2. p (or V) and H (or E): This group includes calculations for adiabatic reactions at constant p (or V), in other words: H (or E) is specified to be the same as that of the reactant system. Such calculations are encountered very frequently in problems involving adiabatic flame temperatures or explosion temperatures obtained from specified flammable or explosive mixtures. Thus, the initial system from which, by irreversible reaction, the equilibrium system is reached, has to be specified as well (both initial composition and temperature (T_0)). Then, the requirement that ΔH (or ΔE) be zero, leads to the (for ideal gases well-known) equations (Lewis and Randall, 1961):

$$\Delta H_r + \int_{T_0}^{T_q} C_p \, dT = 0 \tag{1.6.2}$$

or

$$\Delta E_r + \int_{T_0}^{T_q} C_v \, dT = 0, \tag{1.6.3}$$

where T_q is the temperature reached in the equilibrium composition, C_p (or C_v) is the total heat capacity of the equilibrium mixture, and ΔH_r (or ΔE_r) is the heat effect $(-Q)$ associated with the adiabatic reaction r, at constant p (or V). For ideal systems:

$$\Delta H_r = -Q_p = (\sum_i n_i H_i^\circ)_{\text{products}} - (\sum_i n_i H_i^\circ)_{\text{reactants}}$$

or $\quad \Delta E_r = -Q_v = (\sum_i n_i E_i^\circ)_{\text{products}} - (\sum_i n_i E_i^\circ)_{\text{reactants}},$

where H_i° and E_i° are standard enthalpies and standard energies of all species at the reactant temperature T_0.

For non-adiabatic systems where H is specified, reference to a reactant system makes possible the calculation of ΔH (or ΔE), which then replaces the zeros on the right-hand sides of equations (1.6.2) and (1.6.3).

For non-ideal systems, the relationships (1.6.2) and (1.6.3) are more complex; for their treatment, the interested reader is referred to Beattie (1949) and Brinkley (1956).

One of the simplest methods for computing the equilibria in this group is to estimate T_p or T_v, and to evaluate a new, improved estimate of T_p or T_v from equation (1.6.2) or (1.6.3); this constitutes in effect an outside iterative loop. More elegant and faster are techniques wherein the adiabaticity iterations are implicit in the computational scheme (see e.g. §§2.5 and 4.3).

3. p (or V) and S: This group represents calculations for isentropic (adiabatic, reversible) processes, i.e. for calculations of equilibria in which S is specified to be the same as that for the reactant system from which an equilibrium is derived. Applications are found in expansion processes such as in rocket nozzles, in rock blasting, etc. The requirement that ΔS for the process be zero, leads to the expression

$$S_{T_0} = S_{T_q}, \tag{1.6.4}$$

where, for ideal gases, at both temperatures initial $T = T_0$ and equilibrium $T = T_q$ (see equation 1.2.9):

$$S_T = \sum_i n_i (S_T^\circ)_i - \sum_g n_g R \ln n_g p, \tag{1.6.5}$$

where g denotes that gaseous species only are included, and where the composition i and pressure p differ in states o and q.

As in group (2) above, the simplest procedure is to estimate T_q, then to calculate the chemical equilibrium at p (or V) and T_q, and to obtain a better estimate of T_q from equations (1.6.4) and (1.6.5). However, as in group (2) a more elegant and quicker, implicit procedure can be followed (see §4.3). Again, non-ideal systems are more complex; for their general treatment, see Beattie (1949) and Guggenheim (1952).

2

MATHEMATICAL STRUCTURE OF
THE CHEMICAL EQUILIBRIUM
PROBLEM

2.1 Introduction

At first sight, the number and variety of methods of solving the
chemical equilibrium problem which have so far appeared in the
literature is rather impressive. Closer examination, however, soon
discloses that most of the suggested methods are essentially minor
modifications of earlier work which scarcely differ from their
predecessors even from the point of view of the numerical analysis
involved. This point will be discussed in greater detail in chapter 5.

Although as far as the mathematical content of the problem is
concerned, all approaches must be equivalent, from the numerical
point of view they are not. Two major types of method can be
distinguished. The first of these usually involves the search for
the minimum (or maximum) of a function and will be referred to
as 'optimization' methods (see for instance Anthony and Himmelblau, 1963). The second type of solution reduces to the numerical
solution of a set of simultaneous non-linear equations (e.g.
Scully, 1962). This type will be referred to as the 'non-linear
equation' approach. Methods based on the use of equilibrium
constants normally fall in the non-linear equations group, while
those based on the minimization of the Gibbs free energy often
(but not always) lie in the optimization group.

In allocating methods between the two groups the decision will
be based entirely on the numerical techniques involved. For
example, a method which may originally start out to minimize
the Gibbs free energy, but reduces in fact to the numerical
computation of a set of simultaneous non-linear equations, will
be regarded as a non-linear equation method. Nearly all the
methods encountered in the literature fall into one or other of
the two main groups, which differ sharply in the type of numerical

difficulties encountered. Several subdivisions can be made within the two groups. This point will be treated in the relevant chapters. In this chapter, the underlying algebraic basis of the numerical problem will be discussed. This discussion will be based, to start with, on the behaviour of the simplest system of any practical interest, that of a mixture of reacting ideal gases at constant temperature and pressure. The numerical problem involved in finding the equilibrium composition will be stated for this system in terms of the minimization of the Gibbs free energy of the system. (This form of the problem has been chosen in order to simplify the later parts of the chapter.) We shall then show how the mass action equation form of the problem can be derived from the Gibbs free energy minimization form.

The chapter will continue with a discussion of the additional problems which arise when condensed species are present in the system. (Non-ideal gases are discussed separately in chapters 3, 4 and 5.) This will be followed by an outline of the effects on the problem of allowing such quantities as the temperature and pressure to vary as well as the composition. The chapter will conclude with a treatment of the effect on the equilibrium composition of variation in the parameters of the system, with particular reference to the effect of errors, and with a brief survey of some of the more abstract work on the chemical equilibrium problems.

2.2 The minimization form: *Ideal gases at constant temperature and pressure*

The computation of chemical equilibria can be regarded from the point of view of the actual algebra, as a mixed linear/non-linear optimization subject to linear side conditions. We may conveniently begin with the assumption that the change in the Gibbs free energy for any infinitesimal process in which the amounts of species present may be changed by either the transfer of species to or from a phase or by chemical reaction, is given by (see equation 1.2.23):

$$dG = -S\,dT + V\,dp + \sum_i \mu_i\,dn_i. \qquad (2.2.1)$$

Here G, S, T and p are the Gibbs free energy, the entropy, the temperature and the (total) pressure. μ_i is the partial molal free energy of species number i, and n_i is the number of moles of species number i in the system. If it is assumed that the temperature and pressure are held constant during the process, dT and dp both vanish. If we now make changes in the n_i such that $dn_i = dkn_i$, so that the changes in the n_i are in the same proportion, then, since G is an extensive quantity (Zemansky, 1957), we must have $dG = dkG$. This implies that

$$G = \sum_i \mu_i n_i. \tag{2.2.2}$$

Comparison of equations (2.2.1) and (2.2.2) shows that the chemical potentials are intensive quantities, since if all the n_i are increased in the same proportion at constant T and p, the μ_i must remain unchanged for G to increase in the same rate as the n_i. This invariance property of the μ_i is of the utmost importance in restricting the possible forms that the μ_i may take.

Equation (2.2.2) expresses the Gibbs free energy in terms of the mole numbers n_i, which appear both explicitly and implicitly (in the μ_i) on the right-hand side. The Gibbs free energy is a minimum when the system is at equilibrium (Zemansky, 1957). The basic problem, then, becomes that of finding that set of n_i which makes G a minimum. Two further points must be considered before even the simplest form of equilibrium problem can be fully stated.

First, the mole numbers n_i represent the amount of species number i present (in moles) in the system. Consequently the n_i must all satisfy the condition

$$n_i \geqslant 0 \tag{2.2.3}$$

since negative mole numbers are physically inadmissible. (This innocent seeming side condition is in fact responsible for a large part of the numerical difficulties of the problem, as will be seen below.)

The second, and more familiar, condition arises from the requirements of the law of conservation of mass. Although any given species in a mixture is a variable, the number of atoms of a specific element is not. The set of n_i values which form the

solution, then, must also satisfy mass balance conditions of the
form
$$\sum_i a_{ie} n_i = B_e, \qquad (2.2.4)$$

where N is the number of species present, M is the number of
elements present $(N \geqslant M)$, a_{ie} is the number of atoms of element
e present in species i and B_e is the total number of gram atoms of
element e present (also called the abundance of element e). The
mass balance conditions (2.2.4) are (happily) linear, equality side
conditions, and although more elaborate than conditions (2.2.3)
will be found a good deal less troublesome. The main source of
difficulty is the presence of the inequality in condition (2.2.3).

It is now possible to state formally the *equilibrium problem* for
the system and conditions considered in this section, as *that of
finding that set of n_i values which minimizes*

$$G = \sum_i \mu_i n_i \qquad (2.2.2)$$

subject to the side conditions

$$\sum_i a_{ie} n_i = B_e \qquad (2.2.4)$$

and $\qquad n_i \geqslant 0. \qquad (2.2.3)$

It should be noted that numerical methods have been devised
which minimize (or maximize) functions which may not be
standard thermodynamic functions, but, for example, measures
of the error of an initial estimate of the solution. The numerical
problems involved are usually the same, as will be seen in subsequent chapters. All methods of this form have been classed as
'optimization' methods, although the need for optimization may
stem from numerical rather than from thermodynamic considerations.

2.3 The mass action equation approach

In this section we shall show that the mass action equation form
of the chemical equilibrium problem can be derived directly
from the Gibbs free energy minimization form. The two approaches are, in fact, mathematically equivalent. In practice,
however, the much greater generality of the Gibbs free energy

minimization form makes it more suitable for all but the simplest of systems, as well as more convenient for theoretical discussion such as that of this chapter.

The proof will be based on the use of the method of Lagrangian multipliers which is probably the most important single technique to be employed on the chemical equilibrium problem. It is beyond the scope of this book to give a proof of the Lagrangian multiplier method. For a straightforward (if not unduly rigorous) proof the reader is referred to Zemansky (1957) (see also Appendix I). The more mathematical reader will probably prefer the treatment given by Courant (1957).

To employ the method of Lagrangian multipliers, we form first the function

$$L = \sum_i \mu_i n_i - \sum_e \chi_e (\sum_i a_{ie} n_i - B_e), \qquad (2.3.1)$$

where the χ_e are a set of M so far indeterminate parameters (the Lagrangian multipliers). The extremum of G, for the system of ideal gases only at constant temperature and pressure, (given by equation 2.2.2), subject to the side conditions (2.2.4) is then given by the solution of the set of equations:

$$\frac{\partial L}{\partial n_i} = 0. \qquad (2.3.2)$$

As will be seen below, the condition (2.2.3) is not important for a mixture of ideal gases only.

Substituting for L in (2.3.2) from (2.3.1) leads to the set of simultaneous non-linear equations:

$$\mu_i - \sum_i \chi_e a_{ie} = 0. \qquad (2.3.3)$$

These N equations, together with the M equations (2.2.4) can (at least in theory) be solved for the $(N+M)$ desired values of the n_i and χ_e at equilibrium. In practice, such solutions must be obtained numerically. A substantial proportion of non-linear equation methods attack the problem at this stage by devising numerical methods for the solution of these equations.

The Lagrangian multipliers do not appear in the mass action equation form of the problem at all, and so must be eliminated.

A simple and elegant way of doing this has been described by Brinkley (1946). The species i are divided into two groups, viz. components, denoted by c, and derived species, denoted by j. Then, the mass balance equations (2.2.4) can be rewritten as:

$$\sum_c a_{ce} n_c = B_e - \sum_j a_{je} n_j. \qquad (2.3.4)$$

They can be considered as a set of equations in the n_c, from which explicit expressions for the n_c can be obtained by inversion, leading to:

$$n_c = \sum_e \bar{a}_{ec} B_e - \sum_e \bar{a}_{ec} \sum_j a_{je} n_j, \qquad (2.3.5)$$

where $[\bar{a}_{ec}]$ is the inverse of the matrix $[a_{ce}]$. Then equation (2.3.5) can be simplified to:

$$n_c = q_c - \sum_j a_{cj} n_j. \qquad (2.3.6)$$

Thus, the q_c and a_{cj} are simple linear combinations of the original B_e and a_{je}, although they are no longer necessarily positive. Equations (2.3.6), however, represent precisely the same restrictions on the n_i that equations (2.2.4) do. Systematic algebraic manipulations of the mass balance equations play an important part in several methods of treating the chemical equilibrium problem. The reader is referred in particular to chapter 4 for details of some of these.

If the method of Lagrangian multipliers is used on the problem as before, but employing this time equations (2.3.6) instead of (2.2.4), equations (2.3.3) become:

$$\mu_c - \chi_c = 0 \qquad (2.3.7)$$

and

$$\mu_j - \sum_c \chi_c a_{cj} = 0. \qquad (2.3.8)$$

(The Lagrangian multipliers χ_c are used to distinguish them from the χ_e of equations (2.3.3).) The χ_c can now be eliminated, since, from equations (2.3.7), $\chi_c = \mu_c$, and these can be inserted in equations (2.3.8) to give:

$$\mu_j - \sum_c \mu_c a_{cj} = 0. \qquad (2.3.9)$$

It may be pointed out here that equations (2.3.7) and (2.3.8) lead directly to the phase rule for a chemically reactive system (equation 1.3.12). This can be demonstrated as follows:

Equations (2.3.7) are the equations of phase equilibrium. For each component species c, there are $\Phi - 1$ equalities of the μ_c, and thus a total of $C(\Phi - 1)$ equalities. Also, there are $(N - C)$ equations (2.3.9), applicable in each phase; thus there are $\Phi(N - C)$ such equations. Therefore, the total number of equations is $C(\Phi - 1) + \Phi(N - C) = \Phi N - C$. As for a non-reactive system (see §1.3.2) these equations represent relationships between $N\Phi + 2$ variables (the x_i and two thermodynamic state variables). Since, in addition, there is a relationship $\sum_i x_i = 1$ for each phase, there are a total of $\Phi N - C + \Phi$ equations between $N\Phi + 2$ variables. The variance F is the difference between the number of variables and the number of equations, thus:

$$F = (N\Phi + 2) - (\Phi N - C + \Phi) = C - \Phi + 2. \qquad (2.3.10)$$

Equations (2.3.9) are the equations of reaction equilibrium. Introducing now the assumption that the system being considered consists of ideal gases only, the μ_i have the form (see equation 1.2.26):

$$\mu_i = \mu_i^\circ + RT \ln p_i. \qquad (2.3.11)$$

The μ_i° may be regarded as constants, characteristic of the corresponding species, at a given temperature T. It should be noted that the second term on the right-hand side of equations (2.3.11) implies that the n_i are all non-negative. Thus the use of this form of μ_i implies that the n_i in the solution will satisfy conditions (2.2.3).

Substituting equations (2.3.11) into equations (2.3.9) and separation of the terms containing p_c leads easily to

$$p_j^{-1} \prod_c (p_c)^{a_{cj}} = \exp\left\{ \frac{1}{RT} (\mu_j^0 - \sum_c \mu_c^0 a_{cj}) \right\}. \qquad (2.3.12)$$

Equations (2.3.12), together with equations (2.2.4) can be solved, again numerically, for the equilibrium n_i values. Equations (2.3.12) have the form usual when the mass action equation method is used, although in this case it is normal to write down a series of equations of the form of (2.3.12) from intuitive considerations, based on the 'main' reactions which should be occurring in the system (see e.g. Mingle (1962)) at equilibrium.

The right-hand side of the equations (2.3.12) are in fact the equilibrium constants for the reactions represented in terms of partial pressures by the corresponding left-hand sides: the a_{cj} are seen to be equivalent to the stoichiometric coefficients for these reactions (cf. §4.2).

The discussion of this section has served then not only to show that the two best known approaches to the equilibrium problem are equivalent but also to make clear the essential limitations of the mass action equation approach, which depends on the assumption of μ_i of the form of equations (2.3.11).

Apart from this, since, at least for systems of ideal gases only, *any* M species can be chosen as components, there is present in the problem an underlying ambiguity which has been the subject of some theoretical work (see e.g. Aris, 1965). This ambiguity will hold of course in the derivation of equations (2.3.12), making any rearrangement of the same equations equally correct. Thus the entire concept that some reactions are more 'important' than others at equilibrium would seem of doubtful validity (see §5.1).

2.4   The inclusion of pure condensed species

We have assumed in this chapter so far that the system being considered consists of ideal gases only, so that the form of μ_i used is that given by equation (2.3.11), and that this in effect forces any n_i values obtained to satisfy automatically the side conditions (2.2.3). This assumption is necessary if mass action equations are to be used, and makes even the simplest generalization of the ideal gas system, the addition of solid species, difficult to treat by the mass action equation method. In this section it will become apparent that, except for the simplest systems, methods based on minimization of the Gibbs free energy, or allied quantities, are substantially easier to use for numerical work.

The inclusion of condensed phases requires a modification of the expression (2.3.11) for the chemical potential. The procedure for the inclusion of pure condensed species is deceptively simple. The second term in equation (2.3.11) represents the effect of the mole fraction of a species i; for a pure species i: $x_i = 1$. Also, the chemical potential for a solid or liquid is independent of pressure (Prigogine and Defay, 1954) so that:

$$\mu_i = \mu_i^\circ. \tag{2.4.1}$$

It is sometimes convenient to use one equation instead of two, to express the chemical potential. One can use:

$$\mu_i = \mu_i^\circ + \zeta_i RT \ln p_i, \tag{2.4.2}$$

where $\zeta_i = 0$ for pure condensed species, and $\zeta_i = 1$ for mixed phases or for pure gases (where $p_i = p$). μ_i° is thus a function of T only, for all phases.

For mixed condensed phases, only the pressure term disappears and one obtains:

$$\mu_i = \mu_i^\circ + \zeta_i RT \ln x_i. \tag{2.4.3}$$

The use of chemical potentials of both forms (2.3.11) and (2.4.1) in the same problem has more profound consequences than might at first be supposed. This is because, if gases only are present, no n_i can actually vanish at equilibrium, although some n_i may become very small. The position of the minimum is thus determined essentially by the function G and is a true minimum, in the sense that it occurs at the point where the appropriate set of partial derivatives vanish. In this case, as has already been noted, the inequality conditions (2.2.3) are of only minor importance.

In a system in which for example *only* solid species were present (a case of some importance in the study of alloys, see Kubaschewski and Evans, 1956), the minimization problem would become in fact a linear programming problem, and the position of the minimum would depend principally on the presence of the inequality condition (2.2.3). In this case all but M of the N (solid) species *must* vanish (cf. §1.3.3).

The presence of both pure condensed phases and gases (or linear and non-linear terms in G) makes it uncertain which type of minimum occurs at equilibrium, and it is quite possible for a mixture of both types to arise. The possibility of minima generated by side conditions tends to make computational methods, designed and tested for systems of gases only, break down if solids or pure liquids are included, and conversely pure linear programming methods which have also been tried (White *et al.* (1958)) are very difficult to use. While it might be thought that solids could be incorporated by simply allowing them a partial pressure term of some sort which could be made vanishingly small at the end of the computation, this does not appear to be satisfactory in practice, and in any case is of little help if a method is desired in which, for example, non-ideal gases are to be included.

As a practical example of the type of difficulty which arises consider equations (2.3.7) and (2.3.8). It is clearly essential that only species which are not zero at equilibrium be considered as components, or the process of eliminating the χ_c will break down. Unfortunately, it is not known, at the start of an equilibrium computation with a totally strange system, which solid species (if any) will be present at equilibrium. Yet it is necessary to include as components those species that *are* present at equilibrium. This makes it necessary, if the mass action equation method is used, to prepare a complete set of equations for each possible combination of solids in the final solution, to carry out the full computation for each set, and to find which of the solutions yields the lowest Gibbs free energy (thus in effect minimizing the Gibbs free energy in any case). While this process is not too tedious if only one or two solids are being considered, it very rapidly becomes quite impractical if the number of possible solids is greater, as is shown below.

In the mass action equation methods (see §2.3) equilibrium constants are used to relate the various concentrations in the system. For each added pure condensed species, one more mass action equation (2.3.12), with its appropriate equilibrium constant becomes available for the computation; however, the mass balance equations containing terms for that solid species then become redundant for that calculation, and only serve to give the amounts of the solid species after the computation for the gas phase has been completed. Thus, the procedure for considering condensed species then becomes the following (Storey, 1965):

Those mass balance equations (2.2.4) which include terms in the condensed species are written in the form:

$$\sum_s a_{se} n_s = B_e - \sum_g a_{ge} n_g, \qquad (2.4.4)$$

where the subscripts s and g refer to condensed and gaseous species, respectively. From equations (2.4.4), explicit expressions for the n_s in terms of the n_g can be obtained by inversion, viz:

$$n_s = \sum_e \bar{a}_{es} B_e - \sum_e \bar{a}_{es} \sum_g a_{ge} n_g \qquad (2.4.5)$$

which can be abbreviated to:

$$n_s = B_s - \sum_g a_{sg} n_g. \qquad (2.4.6)$$

Equation (2.4.6) can then be inserted into the relevant equations (2.2.4):

$$\sum_s a_{se}[B_s - \sum_g a_{sg} n_g] = B_e - \sum_g a_{ge} n_g \qquad (2.4.7)$$

and this assumes the simple form:

$$\sum_g b_{gh} n_g = Q_h, \qquad (2.4.8)$$

where $h = 1 \ldots (M-S)$. Thus, by the above procedure, the M mass balance equations have been reduced in number by the number of pure condensed species in the system. Any solution of the system of $(M-S)$ equations (2.4.8) and $(N-M)$ equations (2.3.12) leads to the $N-S$ values of the n_g, and subsequently, with equation (2.4.6), to the S values of the n_s.

The above procedure requires that equations (2.4.4) be solved for the n_s. This is necessary for *each* different selected set of solid species assumed present in the system. In other words: the set of equations (2.4.6) varies as the estimates vary, and the whole solution structure is affected. Furthermore, the maximum number of condensed species that can be considered in the mass action equation methods at any one time is determined by the phase rule, i.e. $S_{\max} = M-1$; this restriction does not apply to the minimization methods.

In general, most successful attempts to devise computational techniques which will allow for both solids and gases in the system have been based either on the direct optimization of quantities such as the Gibbs free energy or, at most, on some stage intermediate between full minimization and the mass action equation form. One method which is ostensibly a minimization method, but is based in fact on equations of the form

(2.3.12) tends to break down if solids are included (Naphtali, 1960, 1961). However, several of the methods to be described in subsequent chapters are independent of the forms of μ_i used.

2.5 The effect of variations in the temperature and pressure

A considerable proportion of the applications of the methods to be described in later chapters lie in the fields of explosives and propellants. Here, it is common to require that either the temperature or the pressure (or both) as well as the composition of the system be variable. Again, important applications are to be found in the design of chemical processes, where it is desired to optimize all the process variables. The subject of chemical process optimization lies beyond the scope of the book. However, much of the material presented here, as well as several of the techniques discussed in chapter 3, (and particularly its first section), will be found of use to those interested in the problem. Although the inclusion of, say, the temperature as a variable is algebraically more complicated than the inclusion of condensed phases, it normally has only a relatively minor effect on the computational techniques to be used.

Consider, for example, a system for which the temperature, but not the pressure, is to be allowed to vary. Those methods which incorporate the addition of the temperature as an independent variable most easily are probably the first-order optimization methods which search for the minimum of G using equation (2.2.1) in the form

$$\frac{dG}{d\lambda} = -S\frac{dT}{d\lambda} + \sum_i \mu_i \frac{dn_i}{d\lambda}, \qquad (2.5.1)$$

where λ is a search parameter of some convenient form. The computer programme written to carry out such a search would be of the same form as one for systems in which the temperature was to be held constant. The only differences would be that the μ_i could no longer be evaluated as functions of the n_i alone but must include the effect of a variation in T, and that a means of calculating the entropy S must be available. If derived thermodynamically from first principles, such expressions are quite complicated in form (Zemansky, 1957), and involve the numerical evaluation of integrals usually based on the specific heat at constant pressure (cf. §1.5). For numerical purposes, simpler approximations are usually used, such as polynomials in T, found by fitting experimental data. The work of Wiederkehr (1962) and Zeleznik and Gordon (1961) has been of considerable practical value in dealing with such problems. Similar considerations apply for those methods based on non-linear equations, where the additional variables must be incorporated into the basic equations, and additional (usually also non-linear) equations must be found to allow the additional variables to be specified.

Variations in T and p can also be incorporated in the mass action equation approach, but in doing so additional equations must be 'tacked on' to the original set. The result is usually too over-specialized to one class of system (see, however, Lu, 1967; van Zeggeren and Storey, 1969) to be of great general interest.

The Gibbs free energy is not, of course, the only function on which optimization approaches can or have been based. It may in some cases be more convenient to make use, for example, of the Helmholtz free energy A, for which

$$dA = -p\,dV - S\,dT + \sum_i \left(\frac{\partial A}{\partial n_i}\right) dn_i \qquad (2.5.2)$$

which attains a minimum at equilibrium, subject to the same side conditions that were used for the Gibbs free energy minimization. The use of thermodynamic quantities other than the Gibbs free energy is usually motivated by convenience. For example, equation (2.5.2) is convenient if the equilibrium is to be found under conditions of constant volume and temperature. The function A is not in fact any more difficult to use than G since

$$\left(\frac{\partial A}{\partial n_i}\right)_{V,\,T,\,n_l} = \left(\frac{\partial G}{\partial n_i}\right)_{p,\,T,\,n_l} = \mu_i.$$

The great bulk of work in the field, however has been done using the Gibbs free energy.

2.6 The effect of changes in parameter on the solution

Relatively little attention has been paid to the problem of finding the effect of changes in the parameters of the problem on the solution once it has been obtained, although some results are available (see e.g. Shapiro, 1964a; Zeleznik and Gordon, 1962, 1968; Neumann, 1962).

In view of the unavoidably questionable accuracy of much of the data available for use in determining the basic parameters needed (e.g. the μ_i°) the effect of experimental and other errors becomes of some importance. The most convenient way of approaching the problem is via the Gibbs free energy by means of a second application of the method of Lagrangian multipliers (Storey, 1965).

In order to illustrate how the problem can be attacked in a practical manner, consider a system consisting of a mixture of perfect gases, at constant temperature and pressure, for which the equilibrium composition has been found. Let it be supposed that there are small errors in the values of the μ_i° that have been employed (see equation 2.3.11). It is then of interest to find what the effect of small (first order) errors, $\delta\mu_i^\circ$ in the μ_i° will be on the n_i values of the solution.

Let it be assumed that the system has been allowed to come to equilibrium at a known set of n_i values n_i' (a prime will be used to denote the value of a parameter in this initial system). Let the values of the μ_i° be changed to $\mu_i^{\circ\prime} + \delta\mu_i^\circ$ and the system again be allowed to come to equilibrium. In the new system, the equilibrium values are taken to be $n_i' + \delta n_i$ where the δn_i are also assumed to be of first order. The change in G in passing from one system (at equilibrium) to the other (also at equilibrium) is given, to second order in the δn_i and $\delta\mu_i^\circ$, by

$$\Delta G = \sum_i \left(\frac{\partial G}{\partial \mu_i^\circ}\right)' \delta\mu_i^\circ + \sum_i \left(\frac{\partial G}{\partial n_i}\right)' \delta n_i + \sum_{i,l} \left(\frac{\partial^2 G}{\partial \mu_i^\circ \partial \mu_l^\circ}\right)' \delta\mu_i^\circ \delta\mu_l^\circ$$

$$+ \sum_{i,l} \left(\frac{\partial^2 G}{\partial \mu_i^\circ \partial n_l}\right)' \delta\mu_i^\circ \delta\mu_l^\circ + \sum_{i,l} \left(\frac{\partial^2 G}{\partial n_i \partial n_l}\right)' \delta n_i \delta n_l, \quad (2.6.1)$$

where the changes δn_i are, as usual, subject to the mass balance conditions:
$$\sum_i a_{ie}\delta n_i = 0 \quad\quad\quad (2.6.2)$$

and it is assumed that the old and new systems both satisfy the inequality conditions (2.2.3). The subscript l is a dummy index for species other than species $i(l = 1 \dots N)$.

Since both the initial system and the final system lie at extrema (minima) of G, the *change* ΔG in G produced by this change in μ_i° must also be an extremum. Thus, since the $\delta\mu_i^\circ$ are fixed, the required values of δn_i, which represent the effect of the changes in the n_i of the original system, must be such as to make ΔG an extremum, subject to condition (2.6.2).

Now the first term on the right-hand side of equation (2.6.1) is constant if the $\delta\mu_i^\circ$ are fixed. The second term must vanish by definition of the original equilibrium. The third term vanishes if equations (2.3.11) are used since $(\partial^2 G/\partial\mu_i^\circ\partial\mu_i^\circ) = 0$. Thus only the last two terms are of interest, and, again using equation (2.3.11), the problem reduces to that of finding that set of δn_i values which makes

$$\sum_i \delta\mu_i^\circ\, \delta n_i + RT\left[\sum_i \frac{(\delta n_i)^2}{n_i'} - \frac{(\delta n)^2}{n'}\right] \qquad (2.6.3)$$

an extremum subject to equation (2.6.2), where

$$n = \sum_i n_i \qquad (2.6.4)$$

and

$$\delta n = \sum_i \delta n_i. \qquad (2.6.5)$$

The method of Lagrangian multipliers, used on the expression (2.6.3) with (2.6.5) and (2.6.2) as side conditions, leads to the conditions

$$\delta\mu_i^\circ + RT\left[\frac{2\delta n_i}{n_i'}\right] - \sum_e \chi_e a_{ie} - \chi_n = 0 \qquad (2.6.6)$$

and

$$-RT\left[\frac{2\delta n}{n'}\right] + \chi_n = 0, \qquad (2.6.7)$$

where δn has been treated temporarily as an independent variable. Equations (2.6.6) can be rearranged to give

$$\delta n_i = \frac{n_i^1}{2RT}\{\sum_e \chi_e a_{ie} + \chi_n - \delta\mu_i^\circ\} \qquad (2.6.8)$$

which is in the form desired, but includes the parameters χ_e and χ_n. These can be found in terms of known quantities as follows:

Summing equation (2.6.8) over i leads to

$$\delta n = \frac{1}{2RT}\{\sum_e \chi_e B_e + \chi_n n' - \sum_i n_i' \delta\mu_i^\circ\} \qquad (2.6.9)$$

and substituting for δn from equation (2.6.7) leads to

$$\sum_e \chi_e B_e = \sum_i n_i' \delta \mu_i^\circ. \qquad (2.6.10)$$

Again, substituting equation (2.6.8) back into the side conditions (2.6.2) leads to

$$\sum_e \chi_e (\sum_i a_{ie} a_{if} n_i') + \chi_n B_f = \sum_i a_{if} n_i' \delta \mu_i^\circ. \qquad (2.6.11)$$

Here B_e is defined by equations (2.2.4), and f is a dummy subscript for $e(f = 1, ..., M)$.

Equations (2.6.10) and (2.6.11) are in fact a set of $M+1$ simultaneous linear equations for the χ_e and χ_n which can be solved fairly easily by numerical means. Once the appropriate χ_e and χ_n are found they may be used in equation (2.6.8) to calculate the appropriate first-order changes in the n_i required.

This technique of applying the method of Lagrangian multipliers a second time appears quite useful for systems in which the form of μ_i used is sufficiently simple, and has, in fact, been used (Storey, 1965) to develop a means of studying the effect of changes in the equilibrium composition of solid-gas systems of systematic changes in another parameter, viz in the elemental abundances (or initial conditions). This application is discussed in chapter 3. Larger changes in the parameters can be treated numerically as a succession of first-order changes, and changes in other parameters, either singly or in combination can also be attacked by this means, although the resulting set of simultaneous equations for the Lagrangian multipliers will only be as simple as (2.6.11) if simple forms for the μ_i are used. In general $N+M$ simultaneous linear equations must be solved instead of the $M+1$ obtained here. Whether or not the use of this technique is faster than direct recomputation of the equilibrium composition under the changed condition unfortunately depends too heavily on the circumstances for any specific rule to be stated.

In cases where solid species are present at equilibrium, but in quantities so small that even a small change in the parameters may cause a change in species actually present at equilibrium, as opposed to a simple change in magnitude, the technique will break down. This difficulty is evident in practice when the matrix of coefficients in the simultaneous equations for the Lagrangian multipliers approaches singularity too closely for easy inversion. In this case recalculation is less troublesome. Fortunately such cases appear relatively rare.

When the above approach is applied to the study of the effect of the elemental abundance (or, in effect, the initial composition of the system) it is possible, by yet another application of the method of Lagrangian multipliers, to develop a technique for searching for optimal initial composition with respect to production of a particular product species. This topic is considered in more detail in chapter 3.

In general terms, little appears to have been published on this problem although Shapiro (1964a) has obtained some useful inequalities and Neumann (1962) has considered errors in the equilibrium constants only, in unfortunately somewhat specialized terms (see also §5.4).

While this book is concerned principally with the *numerical* aspects of the computation of chemical equilibria, it is of some interest to review briefly some of the less frankly numerical work which has appeared in recent years.

Probably the most important single work, with regards to its computational implications is that of Zeleznik and Gordon (1960), who carry out a detailed analytic investigation of what were undoubtedly the three most popular large computer methods in current use, viz., that of Brinkley (1947), Huff *et al.* (1951) and White *et al.* (1958). Zeleznik and Gordon reach the conclusion that no one of the three methods has any significant advantage over the other two. They also show that, under the appropriate conditions, all three methods are guaranteed to converge quadratically. This is of some interest, since as we shall see in later chapters, the three methods are at first sight radically different (see however §5.2).

Another notable worker in the field is Pings (1961, 1963, 1964, 1966) who has been able to derive a number of useful analytical results. In particular he considers, among other problems, the optimization of initial conditions by an analytic technique. Unfortunately his results are not very convenient for numerical work on equilibrium computation for multi-constituent systems, since Pings' treatment applies only to single reaction systems. In a very recent thesis, Lu (1967) has extended much of Pings' work to multiple reaction systems. Lu has investigated displacement of chemical equilibria (see §2.5) as well as the optimization of the equilibrium yield for multi-constituent systems obeying the ideal solution law.

Finally we may note the work of Dantzig and de Haven (1962), considering very large complex biological systems, who derive a compartmentalization technique which for the price of some additional assumptions can greatly reduce the effective size of the system being considered.

Several further contributions to the field of chemical equilibria have been made (see e.g. Aris, 1965) but a discussion of these is felt to lie beyond the scope of this book.

3

METHODS BASED ON OPTIMIZATION
TECHNIQUES

3.1 Introduction

The methods to be discussed in this chapter have been of importance to the study of the chemical equilibrium problem for a relatively short time. They have only become popular (or indeed practical) since the appearance of electronic digital computers. In spite of their greater power, methods of the optimization type often involve more complex and sometimes more voluminous computations than the older non-linear equation or graphical methods, so that the use of a computer is normally essential.

In this chapter, we shall collect the published methods somewhat arbitrarily into three groups. The first of these consists of methods which are essentially *general* optimization techniques which have been applied to the chemical equilibrium problems. These will be referred to as 'general' methods, and include the various pattern search methods, as well as the Linear Programming approach of White *et al.* (1958). These methods tend not to require the assumption that the object function to be minimized is continuous or differentiable, or at least do not rely heavily on this assumption. They are mainly studied in the literature on numerical analysis rather than that of physical chemistry.

The second group of methods we have referred to as 'steepest descent' methods, of first and second order. These methods require the objective function to be both continuous and differentiable to at least first order and have been discussed more frequently in the chemical literature than those of the first group.

Two methods, which, although strictly speaking of the steepest descent type, do not fit in comfortably with those of the second group, have been put in a small group of their own. These are a method of Naphtali's (1960, 1961), which is based internally on a selected set of reactions between species, but minimizes the

Gibbs free energy to reach a solution, and a method due to Storey (1965), which although intended essentially for surveys of the effect of changes in the elemental abundances on a known equilibrium, can be used to calculate equilibria directly. Both these methods are of limited application.

Finally a very brief discussion of work on the general optimization will be presented, largely with a view to providing the worker from the physico-chemical field with an entry into the literature on the more general problem, much of which is of considerable interest.

3.2 General methods

The published literature, in computational, mathematical and other journals on general computational methods, for the solution of the minimization problem is unfortunately far too extensive to be discussed adequately in this book. This will be made clear by the bibliographies of some of the reviews and articles considered in §3.5.

As a result of this huge volume of literature, we shall confine ourselves in this section to discussing those general methods *which have already been applied* to the chemical equilibrium problem. This at once reduces the problem to that of discussing three radically different methods. The first of these may be described as a 'brute force' search, the second is a method which changes the constrained minimization to one without constraints and proceeds to solve the unconstrained form, and the third is a direct application of Linear Programming. Steepest descent methods are, naturally, also 'general' methods in the sense that they have frequently been used on other problems. The steepest descent methods discussed in §3.3, however, make use of the special form of the chemical equilibrium problem.

3.2.1 Pattern search method

In this section we shall base our discussion on the work of Anthony and Himmelblau (1963) who adapt a method due to Hooke and Jeeves (1961) to the chemical equilibrium problem. Hooke and Jeeves' original method was in fact intended for the unconstrained minimization problem. Apart from the work of Anthony and Himmelblau, who perforce had to extend Hooke and Jeeves' work to include constraints in order to use it on the chemical equilibrium problem, it has since been extended by others (see Spang, 1962) and in particular by Glass and Cooper (1965). The latter propose a method specifically designed (as opposed to 'patched up') to deal with the constrained minimization problem.

The 'Direct Search' method, as Hooke and Jeeves have called their method, is in essence quite straightforward. The object is to minimize the value of some (real) function $\Phi(n_i)$ of a set of (real) parameters n_i by searching for the appropriate n_i values. The search, in its original form, is carried out in two stages.

The first stage consists of making one or more 'exploratory' moves, which involve changing the n_i in rotation by small amounts Δn_i and testing to see if any reduction has been made in $\Phi(n_i)$. After the desired number of exploratory moves has been made, and there appears to be no particular preference for any specific number of such moves, the routine enters the second stage. This consists of a 'pattern' move in which each coordinate is changed simultaneously *by an amount proportional to the number of exploratory moves which it has made*, thus allowing the possibility of quite large steps where needed.

Side conditions are taken into account either by simply rejecting unsatisfactory values and returning to the original base point, or by the use of 'penalty functions' i.e. spuriously large values of Φ which should force the search away from a forbidden region in the course of normal operation.

The extension of the Direct Search method by Glass and Cooper (1965) requires a third stage, called by them an alternative routine, the object of which is to extricate the search should it become trapped close to a forbidden region. Since we have no evidence that such an additional stage is necessary for the chemical equilibrium problem, or indeed that it has been used at all, we shall not pursue this particular extension here. A detailed study of the application of such an extension, would, however, be of some interest.

In adapting Hooke and Jeeves' method to the chemical equilibrium problem, Anthony and Himmelblau consider both the minimization of Gibbs free energy and the non-linear equation form of the problem. The latter is converted to the minimization form by requiring that the sum of the squares of the residuals of the mass balance equations be reduced to a minimum (or zero), as in equation (3.2.5) below.

Specifically the Direct Search method is applied as follows:

(i) The minimum of some function such as the Gibbs free energy, G, is found subject to the mass balance conditions (3.2.1) and the non-negative condition (3.2.2)

$$\sum_i a_{ie} n_i = B_e, \qquad (3.2.1)$$

$$n_i \geqslant 0. \qquad (3.2.2)$$

Although this approach is not considered in any detail by Anthony and Himmelblau they suggest that equations (3.2.1) should be solved for M of the n_i in terms of the $N - M$ remaining n_i. The actual search procedure would then be carried out using the $N - M$ independent n_i. The M dependent n_i would be computed at each step, and the combined n_i used to find the current value of G. The new values of n_i can easily be tested at each step to find if the search is moving into a region where any of the n_i are negative. This is then avoided by the use of a penalty function or by simply returning to the original base point.

This particular scheme, if used on a system embodying a large number of solid species, would need exceedingly careful programming, and probably the use of Glass and Cooper's alternative routine to avoid becoming at least inefficient on the simultaneous vanishing of two or more of the solid species during the course of the search. The scheme should, if a variable step size is used, be of roughly the same order of speed as the first-order steepest descent methods described below. Unfortunately, no information has appeared in the literature on these points. Anthony and Himmelblau themselves do not discuss the inclusion of condensed phases, and concentrate their attention on method (ii) below.

(ii) The alternative approach is to use the search technique to solve the mass action equations (with the mass balance equations) as a set of simultaneous non-linear equations. The solution of these equations by iteration and other methods is discussed in detail in chapter 4. The approach described here changes the solution of a set of simultaneous non-linear equations into a minimization problem. Basically, it consists of solving a set of non-linear equations of the form

$$\Phi_j(n_i, \ldots, n_N) = 0 \qquad (3.2.3)$$

together with equations (3.2.1) in the form

$$\Phi_e(n_i, \ldots, n_N) = \sum_i a_{ie} n_i - B_e = 0. \qquad (3.2.4)$$

In this approach, the n_i are adjusted to minimize the function

$$\Phi(n_i) = \sum_j \Phi_j^2(n_i, \ldots, n_N) + \sum_e \Phi_e^2(n_i, \ldots, n_N) \qquad (3.2.5)$$

to zero by means of the search routine. Provided the set of equilibrium equations is correctly chosen, the values of n_i which minimize (3.2.5) will satisfy the conditions (3.2.2) (for systems of gases). In any case it is still necessary to test the n_i to ensure that they satisfy (3.2.2) during the course of the search. If any n_i should become small during the course of the search, the step size should be reduced at once to avoid an incorrect result.

This formulation, via expression (3.2.5), brings into prominence the chief difficulty of solving non-linear equations by minimizing the sum of the squares of the residuals. The problem is, of course, that of deciding the weights to be assigned to the individual terms in forming the sum (3.2.5). Anthony and Himmelblau discuss this point, and suggest weighting the terms according to the accuracy of the thermodynamic data incorporated in the individual equations. This is not unreasonable. However, such a weighting, by grossly distorting the surface on which the search is carried out, can increase the time taken by a search very considerably. It can also aggravate the difficulty with the side conditions that the work of Glass and Cooper was designed to circumvent.

Apart from such difficulties, the choice of weights was found by Anthony and Himmelblau to be of some importance, in that it can have an appreciable effect on the final solution obtained. Several other weights can be suggested, but there does not yet appear to be any work available, even in the general literature, specifically aimed at providing criteria for the choice of a weighting.

Lanczos (1956) has suggested, for simultaneous *linear* equations, a technique for devising weights using the partial derivatives of the Φ; this technique may well be adaptable to non-linear equations.

3.2.2 Unconstrained optimization method

Relatively little (published) work is available discussing application of recent work on the general optimization problem to the computation of chemical equilibria. In this section, however, a recent application by Jones (1967) of a general non-linear programming method, due to Fiacco and McCormick (1963, 1964a, 1964b, 1966, 1967), to the chemical equilibrium problem, will be presented. The numerical method involved has become known as SUMT (Sequential Unconstrained Minimization Technique). SUMT is of particular interest since it is representative of a widespread trend in work on the general minimization problem, in that it converts a constrained problem to an unconstrained problem and then deals with the latter. The discussion given here follows that of Jones (1967).

Briefly, the SUMT method may be described as follows: It is desired to minimize the function (in this case the Gibbs free energy):

$$G = G(n_i, ..., n_N) \qquad (3.2.6)$$

subject to the side conditions

$$\Phi_e = \sum_i a_{ie} n_i - B_e = 0 \qquad (3.2.7)$$

and

$$n_i \geqslant 0. \qquad (3.2.8)$$

The basic SUMT algorithm is then
(i) Form a function

$$Q(n_i, ..., n_N; r_k) = G + r_k \sum_i \frac{1}{n_i} + \frac{1}{\sqrt{r_k}} \sum_e \Phi_e^2. \qquad (3.2.9)$$

(ii) Find the *unconstrained* minimum of Q for a decreasing sequence of positive values of r_k which is (in theory) eventually allowed to tend to zero.

The technique employed by the SUMT algorithm to find the sequence of unconstrained minima is a modified Newton's method. Given a fixed value of r_k, and an initial estimate m_i of the minimizing n_i values, the next estimate of n_i is given by

$$(m_i)_{\text{new}} = m_i - \lambda \sum_l \left[\frac{\partial^2 Q}{\partial m_i \, \partial m_l} \right]^{-1} \left(\frac{\partial Q}{\partial m_l} \right), \qquad (3.2.10)$$

where $[\partial^2 Q/\partial m_i \, \partial m_l]^{-1}$ is an element of the inverse of the matrix with elements $[\partial^2 Q/\partial m_i \partial m_l]$, and λ is a scale parameter, the value of which will depend on the particular method being used. Jones points out that the method of rank annihiliation can be used to reduce considerably the labour involved in inverting the matrix $[\partial^2 Q/\partial m_l \partial m_l]$ in the case of constant temperature and pressure systems, and the reader is referred to his paper (Jones, 1967) for further details. A discussion of Newton's method is given in §5.2.

Once the constrained problem has been put into unconstrained form by the use of equation (3.1.9), a very large body of work is available on the determination of the minimum. In fact, the bulk of the work done so far on the general optimization problem has been on the unconstrained case (see e.g. the review given by Spang, 1962). The key idea in this approach is that of changing the constrained minimization into an unconstrained problem, or more precisely into a succession of unconstrained problems. There is, of course, no special reason why the form of Q given by equation (3.2.9) should be used, although Fiacco and McCormick have found it to work successfully. More especially, there is no reason why techniques other than Newton's method should not be used in solving the resulting unconstrained problem, particularly if, say, a variable temperature was involved, which would make Jones' approach via rank annihilation infeasible.

Among numerical analysts, there appears to be a strong body of opinion that the best way of treating all constrained minimization problems is by conversion into unconstrained problems, principally since this will allow a number of very efficient optimization methods to be used. A detailed discussion of this is beyond the scope of this book. On the face of it, however, the arguments against this point of view would appear to be that it converts a single constrained problem into a *succession* of unconstrained problems, and that if for example equation (3.2.9) is used, the solution obtained can never in practice satisfy the mass balance conditions exactly. There is, unfortunately, not yet sufficient evidence available to shew which point of view is relevant to the special form of the optimization problem involved in the computation of chemical equilibria, so that the approach of this section must for the moment remain attractive but unproven. Further work along these lines may be of considerable value.

3.2.3   Linear programming method

The third of the methods which we discuss here is the application by White *et al.* (1958) of Linear Programming techniques to the chemical equilibrium problem. These authors replace the original (non-linear) chemical equilibrium problem by a linear problem which can be made to approximate the original problem as closely as desired.

The linear problem can then be solved by any of the techniques available for the solution of the linear programming problem, such as the well known Simplex method originated by G. B. Dantzig (see e.g. Koopmans, 1951). Linear Programming methods have also been used by Clasen (1965) to obtain starting values for other methods. A discussion of this can be found in §5.1.1.

The procedure for linearizing the non-linear problem was applied by White and his co-workers to the case of gases only, at constant temperature and pressure. In this case, the Gibbs free energy function becomes

$$G = \sum_i \left\{ c_i + \ln\left(\frac{n_i}{n}\right) \right\} n_i, \qquad (3.2.11)$$

where equation (2.3.11) for the gaseous μ_i has been substituted into equation (2.2.2) thus yielding G for a system at constant temperature and pressure. Note that:

$$c_i = \frac{1}{RT}[\mu_i^\circ + RT \ln p]. \qquad (3.2.12)$$

Equation (3.2.11) is first rearranged as

$$G = \sum_i c_i n_i + n \sum_i \left(\frac{n_i}{n}\right) \ln\left(\frac{n_i}{n}\right). \qquad (3.2.13)$$

Now the functions $\beta_i = x_i \ln x_i,$ (3.2.14)

where $x_i = n_i/n$ (3.2.15) have the form of the solid curve in fig. 3.1. (N.B. Strictly speaking $0 < x_i < 1$ for gases.)

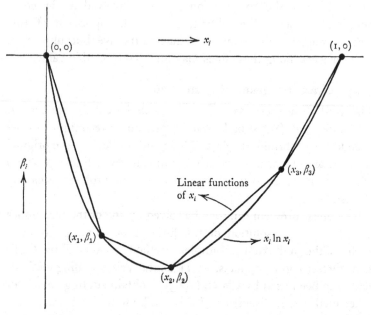

Fig. 3.1. The function $\beta_i = x_i \ln x_i$ and its approximation
(after White, Johnson and Dantzig, 1958).

Each of the functions β_i is then replaced by a piecewise linear approximation as shown in fig. 3.1. If the splines are such that the two functions agree at points (x_p, β_p), $p = 1 \dots P$, then White *et al.* restate the original non-linear problem as that of finding the minimum of

$$\sum_i c_i n_i + \sum_p \beta_p n_{ip} + \sum_p \beta_p n_{2p} + \dots + \sum_p \beta_p n_{Np} \qquad (3.2.16)$$

which is linear in the n_i and the new unknown n_{ip}, subject to

$$\sum_i a_{ie} n_i = B_e, \qquad (3.2.17)$$

$$\sum_i n_i = n, \qquad (3.2.18)$$

$$\sum_p n_{ip} = n, \qquad (3.2.19)$$

$$\sum_p x_p n_{ip} = n_i \qquad (3.2.20)$$

and $$n_i \geqslant 0, \qquad (3.2.21)$$

$$n_{ip} \geqslant 0. \qquad (3.2.22)$$

Equations (3.2.17) and (3.2.18) will be recognized as the usual mass balance equation and the equation defining n, respectively. Equation (3.2.21) is the usual requirement that n_i be non-negative. In addition to the unknown n_i, White *et al.* (1958) have introduced the additional unknowns n_{ip}. These are generated by the linearization process, and allow the function fitted to G to be treated as a single expression instead of having to be broken up piecemeal.

Unfortunately, this process, if attempted directly, very greatly increases the number of unknowns, and leads, if a system of practical interest is being considered, to a formidable linear programming problem. In fact, the solution is beyond the capacity of any but a fairly large digital computer. The originators of the method are aware of this difficulty and recommend approaching the problem in stages. They suggest modifying the normal linear programmes to allow the computation to be carried out, first with a coarse approximation to the function, and then repeated with successively finer approximations *in the region around the immediately preceding solution only* until the desired accuracy is attained. It should be noted however that this formulation of the problem will give zero values for those species present at equilibrium in trace quantities only, so that, if required, they must be found by other means.

On the whole the approach is of considerable theoretical interest. It should also be useful for practical work if only one or two approximate equilibria are needed and the user has access to a large computer for which a linear programme has already been written. For general use, as a method of computing chemical equilibria, the technique is of less interest. Apart from its inherently approximate nature, it is likely to be too expensive, both in computer time and memory space.

3.3. Steepest descent methods

The method of steepest descent has been seriously considered for the numerical solution of the chemical equilibrium problem only since the introduction of the electronic digital computer. The earliest, and still most popular, approach appears to have been that of White *et al.* (1958). Although originally intended for manual use, this method was clearly developed with the use of the digital computer in mind.

In the following discussion of steepest descent methods, we shall distinguish between first-order methods (e.g. Storey and van Zeggeren, 1964), and second-order methods, such as that of White *et al.* (1958).

Briefly, in a first-order method, the computer is simply made to follow a line of steepest slope down the free energy surface, usually in fairly small steps, until it arrives eventually at the minimum. The second-order method consists of two stages. In the first stage, a parabolic surface is fitted to the Gibbs free energy surface at the point the search has reached. Having done this one can either (i) follow a line of steepest descent (as in the first-order approach) but with substantially longer steps or (ii) go directly to the minimum of the fitted parabolic surface as an approximation to the minimum of the Gibbs free energy, using this as the next point in the search. White *et al.* have considered alternative (ii). Alternative (i) does not appear to have been investigated.

3.3.1   First-order method

This method, which has been suggested by Storey and van Zeggeren (1964) for use on small computers, is based on the differential form of the definition of Gibbs free energy (equation 2.2.1) in the form

$$\frac{dG}{d\lambda} = \sum_i \mu_i \frac{dn_i}{d\lambda}, \tag{3.3.1}$$

where it is assumed that the temperature and pressure are constant. The variable λ is a search parameter, which will be accounted for below. The mass balance equations (2.2.4) become, in differential form

$$\sum_i a_{ie} \frac{dn_i}{d\lambda} = 0. \tag{3.3.2}$$

This assumes that the elemental abundances B_e do not change during the computation of the equilibrium and thus implies that *the computation must start from an estimate of the equilibrium which satisfies the mass balance equations.*

In order to avoid practical difficulties during the course of the search as a result of the condition that the n_i be non-negative, the search is carried out in terms of the variable ξ_i, defined by

$$n_i = e^{\xi_i} \tag{3.3.3}$$

in terms of which equations (3.3.1) and (3.3.2) become

$$\frac{dG}{d\lambda} = \sum_i \mu_i n_i \frac{d\xi_i}{d\lambda} \tag{3.3.4}$$

and

$$\sum_i a_{ie} n_i \frac{d\xi_i}{d\lambda} = 0. \tag{3.3.5}$$

Thus, at any given point in the search, the direction of steepest slope is taken to be that which makes $dG/d\lambda$ in equation (3.3.4) an extremum, subject to conditions (3.3.5) and the additional subsidiary condition

$$\sum_i \left(\frac{d\xi_i}{d\lambda}\right)^2 = 1 \tag{3.3.6}$$

since it is the *direction* of steepest slope which is of interest here. The size of a step is adjusted by means of the variable λ.

Applying the method of Lagrangian multipliers leads to the set of equations

$$\mu_i m_i - \chi_\xi \left(\frac{d\xi_i}{d\lambda}\right) - \sum_e \chi_e a_{ie} m_i = 0 \tag{3.3.7}$$

which specify the desired $d\xi_i/d\lambda$ values. Here m_i has been used to denote the value of the n_i at the current point in the search. Equations (3.3.7) can thus be used to find $d\xi_i/d\lambda$ in terms of both known quantities and of the Lagrangian multipliers χ_e and χ_ξ. It is thus necessary to express the χ_e and χ_ξ in terms of known quantities. This is done by multiplying equations (3.3.7) by $a_{if} m_i$ and summing over i. Using equation (3.3.5) this leads to the equations

$$\sum_e \{\sum_i a_{ie} a_{if} m_i^2\} \chi_e = \sum_i a_{if} \mu_i m_i^2, \tag{3.3.8}$$

where f is a dummy subscript for e. Equations (3.3.8) form a set of simultaneous linear equations for the χ_e which can be solved for the χ_e in terms of known quantities only, so that then the $d\xi_i/d\lambda$ can be found from equations (3.3.7), with χ_ξ chosen to satisfy equation (3.3.6).

The computational cycle is then as follows:

(1) Starting from an initial composition m_i, specified by the ξ_i, the appropriate μ_i values are found.

(2) From the μ_i and the m_i, the coefficients in equation (3.3.8) are found.

(3) Equation (3.3.8) is solved for the χ_e. The number of mass balance equations is not usually large. Apart from this, the matrix of coefficients is symmetric and usually quite well conditioned, which makes it possible to use one of a variety of methods for solving for the χ_e, including the term elimination method (Hildebrand, 1956).

(4) Once the χ_e are known, the required $d\xi_i/d\lambda$ are given by

$$\frac{d\xi_i}{d\lambda} = \frac{m_i}{\chi_\xi}[\mu_i - \sum_e \chi_e a_{ie}], \qquad (3.3.9)$$

where χ_ξ is chosen to satisfy equation (3.3.6).

(5) The next set of species concentrations are taken to be specified by

$$(\xi_i)_{\text{new}} = \xi_i + \left(\frac{d\xi_i}{d\lambda}\right)\delta\lambda, \qquad (3.3.10)$$

where $\delta\lambda$ is a step size which controls the rate of progress of the search. The sign of $\delta\lambda$ is chosen to make the search lead to a more negative value of G, i.e. towards its minimum.

This method suffers from two disadvantages. The first of these lies in the tendency of truncation errors to cause the successive estimates to 'drift' to compositions which no longer satisfy the original mass balance conditions. This can be avoided by the use of a smaller step size which, of course, makes the method too slow for hand computation.

On the other hand, the simplicity of the method makes it possible to treat quite large systems in a small computer (given time), and the inclusion of a large proportion of solid species causes no computational difficulties whatsoever. Apart from this, even if the starting composition is chosen very unwisely, it does not appear to suffer from the (numerical) convergence problems associated with methods which attempt to go directly to the minimum in a single step.

The method is thus of considerable value in cases where absolutely nothing is known about a system beforehand, and at worst can be used to provide excellent starting values for faster, if more temperamental, methods such as the RAND method.

The drifting phenomenon could to some extent be cured by employing high-order integration formulae (such as fourth-order Runge–Kutta) instead of the rather crude integration formula used in equation (3.3.10). This approach, however, would probably put the method out of reach of a computer with limited storage capacity. No information on the change in speed or accuracy of the method appears to be available in the literature.

An alternative method of dealing with the drifting phenomenon which works satisfactorily for small systems, but does not seem to have been considered in detail in the literature, is as follows (Storey and van Zeggeren, 1969).

Let the problem be recast as that of finding that set of $\delta\xi_i$ which make

$$\delta G = \sum_i \mu_i m_i \delta\xi_i \qquad (3.3.11)$$

an extremum, subject to

$$\sum_i a_{ie} m_i \delta\xi_i = \Delta B_e, \qquad (3.3.12)$$

where
$$\Delta B_e = B_e - B'_e. \qquad (3.3.13)$$

Here the B_e are the required elemental abundances, and the B'_e are the elemental abundances which actually obtain at the current operating values, m_i of the n_i. This formulation will be seen to be equivalent to that of equations (3.3.4) and (3.3.5), with the exception that ΔB_e, although assumed small, is no longer taken to be zero. The subsidiary condition (3.3.6) is modified to

$$\sum_i (\delta\xi_i)^2 = \sigma^2, \qquad (3.3.14)$$

where the value of σ is now used to control the step size at each cycle.

If the method of Lagrangian multipliers is used as before, equation (3.3.9) becomes

$$\delta\xi_i = \frac{m_i}{\chi_i}\{\mu_i - \sum_e \chi_e a_{ie}\}, \qquad (3.3.15)$$

6

where, however, the χ_e are now given by

$$\sum_e \{\sum_i a_{ie} a_{if} m_i^2\} \chi_e = \sum_i a_{if} \mu_i m_i^2 - \chi_\xi \Delta B_f. \qquad (3.3.16)$$

The χ_e can be written as

$$\chi_e = \eta_e + \chi_\xi \omega_e \qquad (3.3.17)$$

so that the η_e and ω_e are given by

$$\sum_e \{\sum_i a_{ie} a_{if} m_i^2\} \eta_e = \sum_i a_{if} \mu_i m_i^2, \qquad (3.3.18)$$

$$\sum_e \{\sum_i a_{ie} a_{if} m_i^2\} \omega_e = -\Delta B_f. \qquad (3.3.19)$$

It should be noted that if a term elimination method is used, the η_e and ω_e can be found concurrently with only a slight modification in the programme for the basic method.

Equation (3.3.15) can now be rewritten as

$$\delta\xi_i = \frac{1}{\chi_\xi}(D_i - \chi_\xi E_i), \qquad (3.3.20)$$

where

$$D_i = \mu_i m_i - \sum_e \eta_e a_{ie} m_i, \qquad (3.3.21)$$

$$E_i = \sum_e \omega_e a_{ie} m_i. \qquad (3.3.22)$$

Substituting for $\delta\xi_i$ from equation (3.3.20) in the subsidiary side condition (3.3.14) and solving for χ_ξ leads to

$$\chi_\xi = \pm[(\sum_i D_i^2)/(\sigma^2 - \sum_i E_i^2)]^{\frac{1}{2}}. \qquad (3.3.23)$$

Again, it should be noted that there is a lower limit on the value of σ^2 (which controls the step size) which can be used. The sign of χ_ξ is chosen to make the search tend to a more negative value of G. The computational cycle is thus slightly modified. Step 1 involves finding the actual mass balances obtaining, as well as the μ_i. Step 2 must include the determination of the ΔB_e (equation 3.3.13). In step 3 the η_e and ω_e are found using the equations (3.3.18) and (3.3.19). The D_i and E_i are found in step 4 using equations (3.3.21) and (3.3.22). Finally the new ξ_i values are given simply by $\xi_i + \delta\xi_i$, where $\delta\xi_i$ is given by equation (3.3.20) and χ_ξ has been found using equation (3.3.23).

Thus this modification provides a means of reducing or removing the drifting phenomenon, while still leaving the method well within the reach of computers of limited memory. The inclusion of temperature and pressure amongst the unknowns is relatively simple in a first-order method (v. §2.5) although, again, no specific information appears to be available. If the final temperature, for example, were not known, expression (2.5.1) would be used in place of equation (2.6.1), and the normalization (3.3.1) would be extended to include a term $(dT/d\lambda)^2$, suitably weighted to compensate for differences in scale between T and the ξ_i. An additional equation of the form of (3.3.4) would be generated (for $dT/d\lambda$) and the computation of the μ_i would need to take into account the changes which took place in T during the search. In short, with a method of this type the new variable could be included on virtually the same footing as the ξ_i. It should be noted, however, that if the system were to be adiabatic, an additional 'heat balance' side condition, incorporating the heats of reaction involved, would be needed (§1.6).

3.3.2 Second-order methods

The discussion of this section is based on the work of White *et al.* (1958) which treats gaseous systems only, at constant temperature and pressure. This method, known as the 'RAND' method has been extended to include condensed species by Boynton (1960), by Kubert and Stephanou (1960), and by Oliver *et al.* (1962). Its use on a digital computer has been investigated by these authors as well as by, e.g. Marek and Holub (1964).

The basic method presented here considers a mixture of ideal gases only, at constant temperature and pressure. It is assumed that a starting estimate of the equilibrium n_i values, m_i, is known, which satisfies the mass balance condition and is such that $n_i - m_i = \delta m_i$ where δm_i is small (but not necessarily of first order). This implies a fairly good initial estimate of the equilibrium composition.

The Gibbs free energy can then be expanded to second order in δm_i, about m_i, leading to a quadratic approximation, $Q(n_i)$ to $G(n_i)$, viz

$$Q(n_i) = G(m_i) + \sum_i \left(\frac{\partial G}{\partial m_i}\right) \delta m_i + \tfrac{1}{2} \sum_{i,\,l} \left(\frac{\partial^2 G}{\partial m_i\, \partial m_j}\right) \delta m_i\, \delta m_l. \quad (3.3.24)$$

Now, from the definitions (1.2.23) and (1.2.26):

$$\left(\frac{\partial G}{\partial m_i}\right) = \mu_i = \mu_i^\circ + RT \ln \left(\frac{m_i}{m}\right) p, \quad (3.3.25)$$

where

$$m = \sum_i m_i \quad (3.3.26)$$

so that

$$\left(\frac{\partial^2 G}{\partial m_i\, \partial m_l}\right) = RT \left(\frac{\delta_{il}}{m_i} - \frac{1}{m}\right), \quad (3.3.27)$$

where δ_{il} is a Kronecker delta.

Substituting (3.3.25) and (3.3.27) into (3.3.24) and dividing by RT, leads to

$$\frac{Q(n_i)}{RT} = \frac{G(m_i)}{RT} + \sum_i \left[c_i + \ln \frac{m_i}{m}\right] \delta m_i + \tfrac{1}{2} \left[\sum_i \frac{(\delta m_i)^2}{m_i} - \frac{(\delta m)^2}{m}\right],$$

$$(3.3.28)$$

where c_i is given by equation (3.2.12). At a given temperature T, White *et al.* then take as their next estimate of the equilibrium composition that value of n_i which makes Q/RT an extremum, subject to the mass balance conditions

$$\sum_i a_{ie} n_i = B_e \quad (3.3.29)$$

(but not, for the moment, the condition that $n_i \geqslant 0$).

Applying the method of Lagrangian multipliers, one obtains the condition for such an extremum as

$$\left[c_i + \ln \frac{m_i}{m}\right] + \left[\frac{n_i}{m_i} - \frac{n}{m}\right] - \sum_e \chi_e a_{ie} = 0. \quad (3.3.30)$$

These equations, together with equations (3.3.29) and the definition of n, i.e. $n = \sum_i n_i$, form a set of $N+M+1$ linear equations in the unknowns n_i, n and the χ_e (the Langrangian multipliers). These may be solved directly. However, the problem can be simplified somewhat by partially reducing the equations algebraically.

Let equation (3.3.30) be rewritten as

$$n_i = -\Gamma_i + \left(\frac{m_i}{m}\right)n + \sum_e \chi_e a_{ie} m_i, \qquad (3.3.31)$$

where
$$\Gamma_i = m_i \left[c_i + \ln \frac{m_i}{m} \right]. \qquad (3.3.32)$$

Summing equation (3.3.31) over i from 1 to N leads to

$$\sum_e B_e \chi_e = \sum_i \Gamma_i. \qquad (3.3.33)$$

Also, multiplying equation (3.3.31) by a_{if} and summing over i leads to

$$\sum_e \{\sum_i a_{ie} a_{if} m_i\} \chi_e + B_f \left(\frac{n}{m} - 1\right) = \sum_i a_{if} \Gamma_i. \qquad (3.3.34)$$

Equations (3.3.33) and (3.3.34) together form a set of $M+1$ simultaneous linear equations which can be solved for the χ_e and the quantity $((n/m)-1)$ in terms of known quantities. This then provides the values of χ_e and n needed by equation (3.3.31) in order to yield n_i, the next estimate of the minimizing composition.

Since N is often large and M is usually small in cases of interest, the reduction of the simultaneous equations to be solved at each step from $N+M+1$ is well worth while. Indeed, the method would be less than satisfactory for large systems if the reduction were not made.

The above treatment has followed that of White et al. in assuming that the initial values m_i must satisfy the mass balance conditions exactly. This is not, however, necessary (Zeleznik and Gordon, 1960, Levine, 1962). Suppose that, in fact

$$\sum_i a_{ie} m_i = B'_e \neq B_e, \qquad (3.3.35)$$

i.e. the starting composition does not (quite) satisfy the mass balance conditions. In this case, equation (3.3.33) would become

$$\sum_i B'_e \chi_e = \sum_i \Gamma_i \qquad (3.3.36)$$

and equation (3.3.34)

$$\sum_e \{\sum_i a_{ie} a_{if} m_i\} \chi_e + B'_f \left(\frac{n}{m}\right) = \sum_i a_{if} \Gamma_i + B_f. \qquad (3.3.37)$$

Equations (3.3.36) and (3.3.37) can be solved for the χ_e and, in this case, for n/m, as the $M+1$ unknowns. As before, these can be substituted into equations (3.3.31) to find the desired n_i values. From general considerations, it would seem that, if a starting composition which did not satisfy the mass balance conditions were to be used, the discrepancy should be kept small.

The basic computational cycle is then:

(1) Starting from an initial estimate of the equilibrium composition which satisfies the mass balance conditions, the coefficients needed by equations (3.3.34), and the Γ_i are found.

(2) Equations (3.3.33) and (3.3.34) are solved for the χ_e and n. Since the number of simultaneous equations to be treated is normally small, most methods of solution should prove satisfactory.

(3) The new estimates of the equilibrium composition are found from equations (3.3.31).

If a starting composition which does *not* satisfy the mass balance conditions is used, equations (3.3.33) and (3.3.34) are replaced by equations (3.3.36) and (3.3.37), and an extra step to find the current B'_e values is introduced.

One would expect the basic RAND method to suffer from round-off 'drift' away from compositions which satisfy the mass balance equations, in much the same manner as the first-order method described in the previous section. The effect, however, should be much smaller, in view of the smaller number of steps required to reach the equilibrium. The use of equations (3.3.36) and (3.3.37) should further reduce any difficulties due to this particular source.

In the event that a negative value of n_i is obtained during the course of the search, the changes in m_i must be made according to some pre-selected convergence procedure.

One of the disadvantages of the method lies in the effect of species which appear only in very small quantities at equilibrium. If a large number of these are included, the convergence of the RAND method is not satisfactory towards the end of the computation. This can be avoided by systematically dropping species

from a computation as soon as the corresponding n_i value falls
below a predetermined critical level. Unfortunately, if this becomes
necessary early in the computation, the species may regain its importance before the equilibrium is reached, and there is no satisfactory way of detecting this while the computation is in progress.
Finally, White *et al.* point out that an injudicious choice of
atom balances can sometimes lead to a set of mass balance conditions in which the rank C of the a_{ie} matrix is less than M, in
other words: if the number of components is less than the number
of elements (cf. §1.3.3). This again leads to numerical difficulties
during the computation. White and his co-workers suggest that,
in such rare cases, additional species be included until the rank of
the a_{ie} matrix reaches M. This, however, is not sufficient if the
additional species are dropped before the equilibrium is reached.

The inclusion of temperature and pressure as variables can be
carried out in much the same manner as in the first-order case,
although, since this method proceeds to second order, the algebraic
formulation is a good deal more involved, although the spurious
normalization is not required. This problem is discussed briefly
by Oliver, Stephanou and Baier (1962).

The RAND method can quite easily be extended to include
condensed phases (Boynton, 1960; Kubert and Stephanou, 1960;
Oliver *et al.* 1962; and Core *et al.* 1963). The extension is
straightforward and a summary of it will be given here in order
to illustrate how the presence of condensed phases increases the
complexity of the problem. Assuming that the condensed phases
behave ideally, in other words that one can use (see §2.4):

$$\mu_i = \mu_i^\circ + RT \ln \frac{n_i}{n} \qquad (3.3.38)$$

(the pressure term disappears for condensed systems), the
derivation proceeds as in the case of the basic RAND method
described above. Equation (3.3.29) becomes:

$$\sum_i \sum_\phi a_{ie}^\phi n_i^\phi = B_e, \qquad (3.3.39)$$

where
$$B_e = \sum_\phi B_e^\phi \qquad (3.3.40)$$

and
$$\sum_i a_{ie}^\phi n_i^\phi = B_e^\phi. \qquad (3.3.41)$$

The superscript ϕ indicates the number of one of the Φ phases in the system. The result of the Lagrangian differentiation becomes (cf. equation (3.3.31)):

$$n_i^\phi = -\Gamma_i^\phi + \left(\frac{m_i^\phi}{m^\phi}\right)n + \sum_e \chi_e a_{ie}^\phi m_i^\phi. \qquad (3.3.42)$$

By analogy with equation (3.3.37) one then obtains

$$\sum_e \left\{\sum_i \sum_\phi a_{ie}^\phi a_{ij}^\phi m_i^\phi\right\}\chi_e + \sum_\phi B_f'^\phi \frac{m^\phi}{m'^\phi} = \sum_i \sum_\phi a_{ij}^\phi \Gamma_i^\phi - B_{f}. \qquad (3.3.43)$$

Also, by summing equation (3.3.42) over all i, gives for each phase:

$$\sum_e \chi_e B_f'^\phi = \sum_i \Gamma_i^\phi. \qquad (3.3.44)$$

Thus, in this more general case, mainly due to Boynton (1960), there are M equations (3.3.42) and Φ equations (3.3.43), to be solved for the M unknowns χ_e and the Φ unknowns n^ϕ. Comparison of this development with that of White et $al.$ shows immediately that the set of $M+1$ equations derived therein is a special case, i.e. for one phase, the gas phase, only, i.e. for $\Phi = 1$.

For non-ideal systems (gaseous and/or condensed) the method can be quite easily modified, provided a function is available which expresses the interaction between the various species in the systems. The Gibbs free energy of a real (non-ideal) system is given by:

$$G^{\text{real}} = G^{\text{ideal}} + G^E, \qquad (3.3.45)$$

where G^{ideal} is the Gibbs free energy function given without the superscript in equation (3.2.11), and G^E is the excess Gibbs free energy. G^E can be represented by (see equation 1.4.20):

$$G^E = RT \sum_i n_i \ln \gamma_i. \qquad (3.3.46)$$

The only effect of this expression is that equation (3.3.25) will then contain the additional terms $RT \ln \gamma_i$; in that case the further derivation becomes trivial. In many cases expression (3.3.46) is not useful in that no values or functions of $\gamma_i(n_i)$ are available. However, it is then often possible to find forms expressing G^E as a function of composition in such a manner that equation (3.3.27) still can be applied, albeit with only a first-order approximation. An example of such an application was investigated by McGee and Heller (1962), who used the Debye–Hückel theory for the development of a physical model representing a gaseous plasma as an ionic solution. The expression for G^E was sufficiently simple to allow equation (3.3.27) to be used, and to find an expression for $\partial G/\partial m_i$ which conformed to equation (3.3.25).

In spite of its drawbacks, mentioned above, the RAND method has proved popular, and given adequate computing facilities, is certainly attractive and flexible. It is interesting to note, however, that the senior author of the original paper on the method has recently published (White, 1967) a substantially different procedure for the computation of chemical equilibria. This is of the non-linear equation form discussed in chapter 4.

3.4 Miscellaneous methods

In this section, two methods which are difficult to fit in logically into the previous section are discussed. The first of these, due to Naphtali (1959, 1960, 1961) is essentially a first-order minimization of the Gibbs free energy. It differs from the straightforward first-order minimization described in §3.3 in that it makes use of the degree of reaction of a pre-selected group of reactions in the system, and in that it is not explicitly a steepest descent method. Thus, in spirit, it is a mass action equation method, in spite of being a minimization method in appearance. Consequently, it was felt more consistent to include it in this chapter, (since free energy minimization is involved) but not in the section on formal first-order methods.

The second method, due to Storey (1965) was not originally intended as a method for finding equilibria at all. Its purpose was to find how a given equilibrium composition changed when the elemental abundances of the system were changed. The aim of the method was to allow surveys of large numbers of possible initial conditions, with a view to finding the optimal initial composition for the production of a given component in the final equilibrium. The method can and has been used successfully to find individual equilibria. Since it is conceptually more difficult than most of the other methods in the book, and of different form (although essentially a first-order steepest descent) it has been included in this section.*

3.4.1 Gradient method

The treatment of the gradient method given here is based on Naphtali's original paper (Naphtali, 1959), which differs mainly in notation from his subsequent papers (Naphtali, 1960, 1961). More recent work on the method will be discussed below.

In the 'gradient' method (as it is described) Naphtali introduces the variable ϵ_j, which represents the extent of reaction number j (cf. §1.2.6). A change in composition, as the result of the progress of the reactions is given by

$$dn_i = \sum_j \nu_{ij} d\epsilon_j, \qquad (3.4.1)$$

where the number of reactions considered is taken to be the $N-M$ required to form derived species j from selected primary species or components (§2.3). The ν_{ij} are the stoichiometric coefficients, given by

$$\nu_{ij} = \frac{\partial n_i}{\partial \epsilon_j}. \qquad (3.4.2)$$

Assuming now constant temperature and pressure, equation (2.2.1) leads to

$$dG = \sum_i \mu_i dn_i. \qquad (3.4.3)$$

Substituting from equation (3.4.1) leads to

$$dG = \sum_j \{\sum_i \mu_i \nu_{ij}\} d\epsilon_j \qquad (3.4.4)$$

or

$$dG = \sum_j \Delta G_j d\epsilon_j, \qquad (3.4.5)$$

where

$$\Delta G_j = \sum_j \mu_i \nu_{ij} \qquad (3.4.6)$$

is the Gibbs free energy change for reaction j (cf. §1.2.6). By *choosing* $d\epsilon_j$ in the form

$$d\epsilon_j = -\Delta G_j d\lambda, \qquad (3.4.7)$$

---

* A third method has been published very recently (Passy and Wilde, 1968). This method came to the authors' attention too late for inclusion in this book.

where $d\lambda$ is the step size for the search, in a search parameter λ, Naphtali obtains

$$dG = -\sum_{j}(\Delta G_j)^2 d\lambda. \tag{3.4.8}$$

Thus, if $d\lambda$ is chosen positive, the change dG in the Gibbs free energy during a step in the search will always be negative, and will perforce eventually reach the minimum of G required. Using the value for $d\epsilon_j$ given by equation (3.4.7) leads to

$$dn_i = -\sum_{j} \nu_{ij}\Delta G \, d\lambda \tag{3.4.9}$$

or, on insertion of equation (3.4.6)

$$dn_i = -\{\sum_{j}\sum_{l} \nu_{ij}\nu_{lj}\mu_l\}d\lambda \tag{3.4.10}$$

which can be written

$$dn_i = -\{\sum \epsilon_{il}\mu_l\}d\lambda, \tag{3.4.11}$$

where

$$\epsilon_{il} = \sum_{j} \nu_{ij}\nu_{lj}. \tag{3.4.12}$$

Before starting the main cycle of the method, then, the quantities ϵ_{il} are calculated from equations (3.4.2) and (3.4.12) for the reactions chosen. It is also necessary to choose an initial estimate of the equilibrium condition that satisfies the mass balance conditions, although this point is not stressed by Naphtali. The basic computational cycle is then as follows:

(1) The estimates of the μ_i (i.e. the μ_l) are found for the current set of n_i values (m_i).

(2) The quantities dn_i are calculated from equations (3.4.11).

(3) A new set of n_i values are found from $m_i + dn_i$.

Several difficulties are inherent in the method, not all of them noted by Naphtali himself.

First, as pointed out by Naphtali, considerable care should be devoted to the magnitude of $d\lambda$, which must be adjusted to avoid too small or too large a value. This point will be amplified below.

Secondly, although this method avoids the matrix inversions necessary to the methods of the steepest descent type above (§3.3) a certain amount of drift is unavoidable, its severity lying probably somewhere between that of the first- and second-order steepest descent methods described above. (See, however §5.2.)

Finally, the method is not suitable as it stands if a large proportion of solid species are to be included in the system, since
there is no implicit mechanism to stop the corresponding n_i
values from becoming negative. (Unless of course *all* the condensed
phases included are actually present at equilibrium.) This criticism
can of course be aimed at all the search methods which employ n_i
as a basic variable, all of which require that the n_i corresponding
to solid species be kept non-negative by artificial means, which
must slow down the computation.

A most illuminating practical discussion of Naphtali's method
has been given by Snow (1963). Snow found that, even with a
small scale test system based on reactions involving hydrogen
and oxygen and their products, the computation failed at
833 °K. This was due to the presence of several species (OH,
H and O radicals) at equilibrium in very small quantities only.
The computer apparently became involved in the reactions of
these species to such an extent that it did not complete the 'main'
reactions.

Snow overcomes this difficulty by introducing weighting factors
w_j, which have very much the same effect as the transformation
(3.3.3) in the first-order steepest descent search method. Instead
of choosing $d\epsilon_j$ as in equation (3.4.7), Snow selects:

$$d\epsilon_j = -\Delta G_j w_j d\lambda, \qquad (3.4.13)$$

where w_j is made equal to the smallest concentration of any
species involved in the jth reaction. This avoids the difficulty
with very small concentrations, and should also make the treatment of solid species more satisfactory. The price is, of course, a
slowing down of the computation.

The introduction of the w_j, however, leads to a further difficulty.
If reactions of the form $A \rightleftharpoons B$, $B \rightleftharpoons C$ and $2B \rightleftharpoons B_2$, are selected,
and the last is displaced strongly to the right during the com-
putation, then the concentration of B will be small. In this case,
a small w_j will be generated for all three reactions, and the computation will be slowed down unnecessarily. To counter this,
Snow points out that there is in fact no reason why the number
of reactions considered need be confined to the $N-M$ that are
necessary (and independent), and includes additional reactions to

provide alternative means for changes of suitable size to take place. A similar ingenious device was proposed by Goldwasser (1959) for a completely different method (§4.5.2).

Since Naphtali's method finds, in effect, the first derivative specifying the direction the search should follow, as does the method of §3.3.1, it would be of interest to discover what effect higher order integration methods would have. A modification, such as Snow's, which increases the power of this very compact method in the treatment of both solids and gases would be of considerable interest. It is to be hoped that further work will be done on this problem.

3.4.2 Survey method

This technique was originally developed (Storey, 1965) with a view to calculating the *change* in a known equilibrium composition *directly* from a given change in the elemental abundances, i.e. without having to go through the processes normally necessary for calculating equilibria for new systems. The application of the technique is to extended surveys, in which it is desired to find how the equilibrium composition of a given initial system changes as the values of the elemental abundances are changed over certain ranges. Such surveys are often required in order to find that initial (non-equilibrium) mixture which will maximize the yield of some species of interest at equilibrium. As a result of the assumption that the system consists of ideal solids and gases, it is possible to use experimentally derived equilibria of undocumented systems as the starting point of a survey, since as will be shown below, the μ_i° values are not required. (Values of the μ_i° are, however, required for adiabatic reaction systems (van Zeggeren and Storey, 1969).)

The same technique can also be used to calculate one single equilibrium composition for a system at constant temperature and pressure.

The properties of the method in this application are somewhat similar to those of the methods described in §§3.3.1 and 3.4.1. Although as a means for calculating single equilibria, the technique has fairly severe limitations, and for this purpose does require a knowledge of the μ_i° values of the system, it is substantially faster in operation than the method described in §3.3.1.

The basic theory is similar to that discussed in §2.6. Consider a system which has been allowed to come to equilibrium at constant temperature and pressure. In the following this system will be called the original system. The mole numbers n_i will then be such as to minimize the Gibbs free energy subject to the usual mass balance conditions (equation 2.2.4). Suppose now that a small change is made in the mass balances, i.e. that the B_e are altered to $B_e + \delta B_e$ where δB_e is of first order only, and that this new system is allowed to come to equilibrium once again. In the following this new system will be called the changed system. The change in the B_e values will be reflected on changes δn_i in the equilibrium values of the mole numbers n_i. These changes δn_i must satisfy the equations

$$\sum_i a_{ie} \delta n_i = \delta B_e \qquad (3.4.14)$$

which are obtained by subtracting the mass balance equations of the original system from those of the changed system.

In the special case where only pure condensed species are included in the systems the μ_i have the form of equation (2.4.1), and only as many (condensed) species can be present at equilibrium as there are elements present in the system ($N = M$). This is the pure linear case, which must be solved by linear programming techniques and in which the 'minimum' is governed entirely by the side conditions and not by the vanishing of a set of first derivatives. In this case, equations (3.4.14) form a set of simultaneous linear equations which can be solved directly for the required δn_i. (It should be noted that this assumes that the changes in the B_e are small enough to cause changes in the *size* of the n_i only and not in the actual species present at equilibrium. This point will be raised again later in this section.)

In the more general case, when both solids and gases are present, the technique described in §2.6 must be employed, based on the assumption that if G is a minimum in both the original and the changed system, then the change ΔG in G produced by the changes in the B_e must be an extremum.

At constant T and p, ΔG can be expressed as a Taylor's series in the δn_i, viz:

$$\Delta G = \sum_i \left(\frac{\partial G}{\partial n_i}\right)' \delta n_i + \sum_{i,l} \left(\frac{\partial^2 G}{\partial n_i \partial n_l}\right)' \delta n_i \delta n_l \qquad (3.4.15)$$

to second order in the δn_i. The first-order terms in equation (3.4.15) must vanish by definition, since the original system is already at equilibrium. (Here a primed quantity is evaluated using the equilibrium n_i values of the original system.) Thus, making use of the fact that

$$\left(\frac{\partial G}{\partial n_i}\right)_{p,T,n_l} = \mu_i \qquad (3.4.16)$$

the problem then becomes that of finding that set of δn_i which make

$$\tfrac{1}{2}\sum_{i,l} \left(\frac{\partial \mu_i}{\partial n_l}\right)' \delta n_i \delta n_l \qquad (3.4.17)$$

an extremum, subject to the conditions (3.4.14). This set of δn_i will then specify the change in the equilibrium composition that has taken place as a result of the changes in the B_e of the original system.

In order to make the technique practical, it is now necessary to assume that the gases and solids have μ_i of the form of equations (2.3.11) and (2.4.1) respectively, i.e. that

$$\left.\begin{aligned}
\frac{\mu_i}{RT} &= c_i + \ln\left(\frac{n_i}{n}\right) \quad \text{for gases,} \\[2mm]
\frac{\mu_i}{RT} &= \frac{\mu_i^\circ}{RT} \quad \text{for solids,}
\end{aligned}\right\} \qquad (3.4.18)$$

where c_i is given by equation (3.2.12); the μ_i° are constants at a given temperature and pressure, and n is the sum of the mole numbers of the gaseous species. In this case

$$\left.\begin{aligned}
\frac{\partial \mu_i}{\partial n_i} &= 0 \quad \text{if either } i \text{ or } l \text{ refers to a solid,} \\[2mm]
\text{and} \quad \frac{\partial \mu_i}{\partial n_i} &= RT\left[\frac{\delta_{il}}{n_i} - \frac{1}{n}\right] \quad \text{if both } i \text{ and } l \text{ refer to gases.}
\end{aligned}\right\} \qquad (3.4.19)$$

δ_{il} is again the Kronecker delta. It is useful, at this stage, to introduce two new subscripts, g and s, to indicate the distinction between gaseous (g) and condensed (s) species. If there are S pure condensed species present ($s = 1, ..., S$), the number of gaseous species is $N-S$ ($g = 1, ..., N-S$). Substituting equations (3.4.19) into the expression (3.4.17) then reduces the problem to that of finding that set of δn_i which make

$$\sum_g \frac{(\delta n_g)^2}{n_g'} - \frac{(\delta n)^2}{n'} \tag{3.4.20}$$

an extremum, subject to

$$\sum_i a_{ie}\delta n_i = \delta B_e \tag{3.4.21}$$

and

$$\sum_g \delta n_g = \delta n. \tag{3.4.22}$$

(N.B. The factor RT has been dropped from expression (3.4.20).)

If solid species are present, not all of equations (3.4.21) represent side conditions, since the δn_i corresponding to condensed species (represented by δn_s) do not appear in the expression to be minimized. Thus it is necessary to isolate the relevant side conditions from the set (3.4.21). The most straightforward way of doing this is to solve equations (3.4.21) for the δn_s:

$$\delta n_s = \delta q_s - \sum_g a_{sg}\delta n_g, \tag{3.4.23}$$

where δq_s and a_{sg} are linear functions of the δB_e and a_{ie}. Equation (3.4.23) can then be substituted back into the remaining equations of set (3.4.21) to obtain $M-S$ equations relating the gaseous species only, viz:

$$\sum_g b_{gh}\delta n_g = \delta Q_h \tag{3.4.24}$$

where $h = 1, ..., M-S$. The b_{gh} and δQ_h are linear functions of the δB_e and a_{ie}. Thus, if condensed species are present, the desired δn_g are those which make expression (3.4.20) an extremum, subject to conditions (3.4.22) and (3.4.24). The method of Lagrangian multipliers leads at once to

$$\delta n_g = \left(\frac{\delta n}{n'}\right)n_g' + \sum_h \chi_h b_{gh} n_g' \tag{3.4.25}$$

for each (gaseous) δn_g. (The δn_s can be found from equations (3.4.23) once the δn_g are known.) However, in equations (3.4.25) $(M-S+1)$ unknown quantities appear. These are the $M-S$

values of χ_h (the Langrangian multipliers) and δn (the change in the sum of the n_g) which must be found in terms of known quantities.

Summing equations (3.4.25) over g and using equation (3.4.22) leads to

$$\sum_h \{\sum_g b_{gh} n_g'\} \chi_h = 0. \tag{3.4.26}$$

Substituting equation (3.4.25) into the side conditions (3.4.24) leads to

$$\sum_h \{\sum_g b_{gh} b_{gh'} n_g'\} \chi_h = \delta Q_{h'} - \left(\frac{\delta n}{n'}\right)\{\sum_g b_{gh'} n_g'\}, \tag{3.4.27}$$

where h' is a dummy subscript for h. Define now the matrix

$$B = (\sum_g b_{gh} b_{gh'} n_g') \tag{3.4.28}$$

and the matrix I as its inverse. The equation (3.4.26) can then be solved for the χ_h in terms of δn as

$$\chi_h = \sum_{h'} I_{h'h} \delta Q_{h'} - \sum_{h'} I_{h'h} \{\sum_g b_{gh'} n_g'\} \left(\frac{\delta n}{n'}\right). \tag{3.4.29}$$

If the maximum number of solids allowed by the phase rule ($S_{max} = M - 1$, see equation (1.3.20)) are present, B and I consist of a single element ($h = h' = 1$) and the problem becomes particularly simple. In any case, B is at most of order M. Finally, substituting for χ_h into equation (3.4.26) leads to

$$\left(\sum_{h,h'} \{\sum_g b_{gh} n_g'\} I_{h'h} \{\sum_g b_{gh'} n_g'\}\right) \left(\frac{\delta n}{n'}\right) = \sum_{h,h'} (\sum_g b_{gh} n_g') I_{h'h} \delta Q_{h'}. \tag{3.4.30}$$

Equation (3.4.30) thus allows δn to be found in terms of known quantities. Once δn is known it can be used in equation (3.4.29) to find the χ_h. Hence the values of the δn_i which reflect the changes in the initial equilibrium n_i values of the system due to the changes in the mass balances are found from equations (3.4.25) and (3.4.23).

There are two modes in which the above results may be used:

The first of these is that in which, given an initial equilibrium composition, the variation of the equilibrium over a range of B_e values can be found by changing the B_e in a succession of small steps. Since each such step produces a new equilibrium composition, and takes little longer to compute by the above method than a step in, say, the steepest descent method, the saving in time, if a large number of mass balances are to be considered, is substantial. Apart from this (and provided the assumption of ideality in the species is acceptable) the quantities μ_i° do not appear in the equations used above to calculate the δn_i. Hence, by starting from an experimentally obtained initial equilibrium, it is possible to carry out surveys of systems containing species for which the μ_i° are not known.

The second mode of use is that in which the results of this section are used to compute the equilibrium composition of a specified system. To do this an initial estimate of the equilibrium composition is made *which satisfies equation* (2.3.3) (i.e. the equilibrium conditions). Such an initial estimate is unlikely, in general, to satisfy the required mass balance conditions as well. However, such an initial estimate will be a genuine equilibrium condition, even if it has the wrong elemental abundances B_e (Horn and Schüller, 1957). In view of this, since the desired mass balances are known and the initial (incorrect) mass balances are easily found, the survey technique can simply be used to find the change in the initial (estimated) equilibrium when the initial mass balances are changed to the desired values. The B_e are changed in predetermined steps of the appropriate size. This technique has thus one property possessed by no other method of computing equilibrium compositions, in that the user knows in advance how many steps the computation will take. This is frequently extremely useful from the computational point of view.

The survey technique suffers from some disadvantages, apart from the fact that it is algebraically more complicated than the other methods described in this chapter. First, it suffers from the roundoff drift (as do several of the other methods discussed in earlier sections); in this case successive steps can drift away from solutions that satisfy the required mass action conditions. Secondly, if during the course of an extensive survey, a gaseous species becomes very small, the step size must be reduced considerably to retain sufficient accuracy, which slows down the computation. In this case, as in the RAND method, it is advisable to drop this species from the system and restart the survey at this point. The survey must also be restarted if a solid species vanishes. Finally it may sometimes be found that the survey will suddenly start to drift considerably at a particular point in the survey, regardless of step size. This has been interpreted as being due to a sudden increase in importance of a species which has been omitted from the system, and which should be included at this point. Unfortunately, the species to be added must be decided from other considerations (see §5.1).

Although, as a result of the greater complexity of finding starting compositions which satisfy the mass action equations rather than the mass balance conditions, this method is more awkward to use for the computation of single equilibria than, say, the RAND method, there appears to be no other method currently available for the calculation of large numbers of equilibria in a single computation.

3.5 The general literature on optimization

It is always difficult for the worker in one field to find a suitable entry point into the literature of another. This is particularly true when the new field is as scattered as the study of general optimization techniques. Probably the most useful single entry into the computational side are the permuted and subject indices to Computing Reviews, published by the Association for Computing Machinery (Finerman and Revens, 1964, 1966, 1967), and of course Computing Reviews itself. This index provides an extensive coverage, but is not, naturally, all inclusive. (In fact, several of the methods described in this book, which have general applicability, and appeared in the chemical literature, have not been included.) For the more engineering oriented reader reference may be made to a very lucid series of articles by Boas (1963).

A study of current work in the optimization field soon shows that the considerable majority of workers have concentrated on the unconstrained problem. However, unconstrained methods are still of interest for the chemical equilibrium problem, as was seen in §3.2.2. The best known individual methods not previously mentioned appear to be the 'Gradient Projection' method due to

Rosen (1960), and Davidon's (1959) 'Variable Metric' method, later modified by Fletcher and Powell (1963). The gradient projection method is essentially a steepest descent method with orthogonal projection of the gradient into a linear manifold approximating any constraints encountered, and although it appears to deal with constraints most efficiently, lacks the speed of convergence of second-order methods. Fletcher and Powell's method, however, provides an efficient method of treating the unconstrained case. (A recent attempt has been made by Goldfarb and Lapidus (1967) to combine the two methods.)

An excellent (and well annotated) review of earlier work in the general field is given by Spang (1962). More recently Fletcher (1965) has reviewed those methods which do not calculate derivatives (and which do not involve side conditions). This is less well annotated than Spang's paper, but provides some useful references in this area. The reader is also referred to the book by Wilde and Beightler (1967) for a most useful presentation of the field.

Finally, the reader is referred to the collection of papers edited by Lavi and Vogl (1966) and to the recent bibliography by Leon (1965).

It is admittedly unlikely that a general method, applied to the equilibrium problem, will be superior to a method developed to take advantages of the special properties of the problem. It may well be of value, however, to know which of the general methods is most suitable for the solution of the equilibrium problem (or will work at all), so that a single programme could be used for both. As usual, there does not appear to be any information currently available on this question.

4

METHODS BASED ON THE SOLUTION OF NON-LINEAR EQUATIONS

4.1 Introduction

As was discussed in chapter 2, one of the two groups of methods for computing chemical equilibria treats the problem by solving the set of non-linear equations formed from the mass action and mass balance expressions. Many different methods of solution of such sets of non-linear equations have been published. The first such method was discussed in detail by Kandiner and Brinkley (1950), based on the fundamental method described rather summarily by Brinkley in 1947. The method is presented in §4.2 below. Many other methods belonging in this category are variations of Brinkley's general method, but they differ in their treatment, often so much, that they merit separate attention in later sections of this chapter (4.3 and 4.4). Some of the methods belonging in the non-linear equations category are unique in that they do not use a many-variable iteration approach, which normally requires matrix inversion, but reduce the problem to that of solving a set of single-variable equations by an iteration method which does not require explicit matrix inversion (§§4.5 and 4.6). A number of methods have been proposed which do not belong to any of the categories described above, which are of limited applicability, but which are useful for quick hand calculation, for systems of limited complexity (§4.7). Finally, a discussion of the concept of element potentials and the (so far non-linear) methods based on this concept is presented (§4.8). Although a case can also be made out for the earlier part of §4.8 to be included in chapter 3, rather than separate the section it is presented as a single unit here.

4.2 Brinkley's method

4.2.1 Principle of the method

In chapter 1 it is shown that, for a composition consisting of M elements and N chemical species the number of components is equal to the rank C ($\leqslant M$) of the matrix of the stoichiometric coefficients (a_{ie}). It is also shown in chapter 1 that usually, though not necessarily, $C = M$. It is assumed throughout this chapter that indeed $C = M$. The component species denoted by c may be represented by the linearly independent formula vectors α_c:

$$\alpha_c = (a_{c1}, \ldots, a_{ce}, \ldots, a_{cM}). \tag{4.2.1}$$

Thus, in the usual case, there are $N - M$ species i, denoted by j, whose formula vectors α_j are linearly dependent vectors, expressible as linear combinations of the independent vectors α_c. Throughout this chapter, standard summation conventions will be used rather than the more elegant tensor notation proposed by Aris (1965) (cf. §1.3.2). Thus:

$$\sum_c \nu_{jc}\alpha_c = \alpha_j. \tag{4.2.2}$$

The species j are commonly called the 'derived' species, owing to the fact that they can be derived from the component species c through the transformation (4.2.2). Equations (4.2.2) correspond to reaction equations between species $\mathscr{A}_c$ to form species $\mathscr{A}_j$:

$$\sum_c \nu_{jc}\mathscr{A}_c = \mathscr{A}_j. \tag{4.2.3}$$

Thus there is one such reaction equation for each species $\mathscr{A}_j$; the ν_{jc} are the stoichiometric coefficients for reaction j, for which the chemical equilibrium constant is given, according to §1.2.6, by the mass action equation:

$$K_{pj} = \frac{p_j}{\prod\limits_c (p_c)^{\nu_{jc}}}, \tag{4.2.4}$$

where the p_j and p_c are the partial pressures of species j and c, respectively. At a given temperature T, the values of the K_{pj} can be obtained from tabulated values of standard chemical potentials, μ_j° and μ_c° (equation 1.2.33), viz:

$$-RT \ln K_{pj} = \mu_j^\circ - \sum_c \nu_{jc} \mu_c^\circ. \qquad (4.2.5)$$

The main approach described by Brinkley (1947) and by Kandiner and Brinkley (1950) is as follows for homogeneous systems: the mass (or atom) balance equations (see chapter 2, equation 2.2.4), are written in the form:

$$\sum_c a_{ce} n_c = B_e - \sum_j a_{je} n_j. \qquad (4.2.6)$$

Explicit expressions for the n_c can be obtained by inversion, leading to:

$$n_c = \sum_e \bar{a}_{ec} B_e - \sum_j \bar{\nu}_{cj} n_j. \qquad (4.2.7)$$

Brinkley introduces the abbreviation:

$$q_c = \sum_e \bar{a}_{ec} B_e \qquad (4.2.8)$$

so that, on substitution of (4.2.8) into (4.2.7):

$$n_c = q_c - \sum_j \nu_{cj} n_j. \qquad (4.2.9)$$

A physical interpretation of the q_c may be inferred by noting that they represent the composition of the system if all the derived species j are absent. Note that $\bar{a}_{ec} = a_{ce}$ are the (ordinary) stoichiometric coefficients denoting the number of atoms of element e in component c, whereas $\nu_{cj} = \nu_{jc}$ are component coefficients, denoting the number of component formulae c making up the jth species.

The set of equations (4.2.4) and (4.2.9) can be used for the evaluation of the equilibrium composition, in a number of different ways. In the initial treatment, by Kandiner and Brinkley (1950), all calculations were made through the mole numbers, n_j and n_c, and the expression (4.2.4) was first converted from an expression in the p_j, p_c to one in n_j, n_c, viz:

$$n_j = K_{pj} \left(\frac{p}{n}\right)^{(\sum_c \nu_{jc} - 1)} \prod_c n_c^{\nu_{jc}}. \qquad (4.2.10)$$

Thus, equations (4.2.9) and (4.2.10) can be used in the following basic algorithm for ideal gas mixtures at equilibrium.

(a) The q_c are evaluated from the defining equation (4.2.8).

(b) The n_c are evaluated from equation (4.2.9), wherein initially the numbers of moles of the derived species, n_j are set equal to zero.

(c) The n_j are computed from the n_c with equation (4.2.10).

(d) The program returns to step (b) to recalculate the n_c values; this is continued until all the n_c values are accurate to within a prescribed tolerance value τ.

The above algorithm represents a purely iterative approach which, if convergent, is the simplest method for solving the set of equations (4.2.9) and (4.2.10). However, this simple method is usually only convergent when all mole numbers n_c are large compared to all mole numbers n_j. In most cases encountered in practice, purely iterative methods can seldom be used successfully, and alternative methods of iteration have to be found. The secant method can be used to advantage in some cases. The Newton–Raphson method can often be used with much more success; it is described in some detail in §4.2.2. Before proceeding with its description, it is fruitful, at this stage, to consider some improvements in the basic method, which will avoid some of the difficulties which are sometimes encountered. These difficulties are: (a) an extra equation:

$$\sum_j n_j + \sum_c n_c = n \qquad (4.2.11)$$

has to be solved at each stage of the computation, because the value of n is required in equation (4.2.10); (b) the initial estimates have to obey the atom balance equations; this can be done with equation (4.2.8), but matrix inversion is required; furthermore, drifting can occur due to round-off errors, and the equations (4.2.8) do not provide automatic drift correction; and (c) negative mole numbers n_c are sometimes calculated from equation (4.2.9), because of over-correction caused by poor estimates. Difficulties (a) and (b) can be overcome by introducing the following modification:

The atom balance equations are (see equation 4.2.6):

$$\sum_j a_{je}\, n_j + \sum_c a_{ce}\, n_c = B_e \qquad (4.2.6)$$

for all elements e. One can select an arbitrary reference element, $e = r$, for which:

$$\sum_j a_{jr} n_j + \sum_c a_{cr} n_c = B_r. \qquad (4.2.12)$$

Introduction of atomic ratios, $R_e = B_e/B_r$, and combining equations (4.2.6) and (4.2.12) leads to:

$$\sum_j n_j(a_{je} - R_e a_{jr}) + \sum_c n_c(a_{ce} - R_e a_{cr}) = 0. \qquad (4.2.13)$$

The left-hand side can be divided by n, leading to the set of $M - 1$ equations in the mole fractions:

$$\sum_j x_j(a_{je} - R_e a_{jr}) + \sum_c x_c(a_{ce} - R_e a_{cr}) = 0. \qquad (4.2.14)$$

In addition, there is an auxiliary equation replacing (4.2.11), viz:

$$\sum_j x_j + \sum_c x_c = 1. \qquad (4.2.15)$$

The set of M equations (4.2.14) and (4.2.15) can be inverted, as was done for equations (4.2.6), to give:

$$x_c = s_c - \sum_j \bar{\nu}_{jc} x_j, \qquad (4.2.16)$$

where the s_c and $\bar{\nu}_{jc}$ are linear functions of the R_e and a_{ie}. Thus, the x_c can be corrected in a manner very similar to that described above for the n_c, in a purely iterative procedure, using only M, instead of $M + 1$, equations. The solution procedure requires a rearrangement of the $N - M$ equations (4.2.10) in terms of mole fractions, viz:

$$x_j = K_{pj} p^{(\sum_c \nu_{jc} - 1)} \prod_c x_c^{\nu_{jc}}. \qquad (4.2.17)$$

The above-mentioned use of mole fractions rather than that of mole numbers, was mentioned in Brinkley's original paper (Brinkley, 1947) but not used effectively in his subsequent papers on computational procedure. The effectiveness of using mole fractions resides is the fact that, in a Newton–Raphson solution of the equations describing the equilibrium, the inversion of equations (4.2.14) and (4.2.15) is not required *a priori*, i.e. the initial estimates do not have to obey the atom balance equations. In effect, any set of estimated values of the x_j can be automatically corrected to make the next set obey these equations. The reason for this is that equation (4.2.15) uses a defined constant (1), rather than an unknown (n). Furthermore, mole fractions are easier to estimate than mole numbers. The effective use of mole fractions is described in some detail by a number of authors, viz Krieger and White (1948), Winternitz (1949), Sachsel *et al.* (1949) and Martin and Yachter (1951). A number of authors (Horn and Schüller, 1957; Weinberg, 1957; Chu, 1958) use partial pressures instead of mole fractions; this is equivalent to using equations (4.2.14) and (4.2.15), each multiplied by a factor p. There are, however, no numerical advantages in using p_j rather than x_j. The reduction of M equations (4.2.9) to $M-1$ equations (4.2.15) can only be effected for systems wherein pressure and temperature are specified. For systems in which volume and temperature are specified it is numerically advantageous to use mole numbers, because the ratio p/n in equation (4.2.10) is then specified, at least for ideal gases, by the equation of state: $p/n = RT/V$. Thus, for such systems, the auxiliary equation (4.2.11) is not required.

The above-mentioned difficulty (*c*) of calculated negative mole numbers (or mole fractions) can be overcome by the use of logarithmic variables. The credit for this suggested improvement is due to Krieger and White (1948) who introduced logarithms of mole fractions as variables. A very clear description of the best variation of the Brinkley–Krieger–White modification is given by Neumann (1966). The advantages are best illustrated for the iteration method which uses the Newton–Raphson technique for solving sets of non-linear equations.

4.2.2   Newton–Raphson solution

The most popular method employed for solving the sets of equations (4.2.14), (4.2.15) and (4.2.17) is the Newton–Raphson method. It is a general method for solving simultaneous non-linear equations (Hildebrand, 1956). Its application to the equilibrium problem is as follows:

The solution (x_j, x_c) has to obey equations (4.2.14) and (4.2.15). For the estimated solution (y_j, y_c) these equations will in general not be satisfied; thus there are discrepancies which are functions, F_k, of the estimates:

$$F_k = \sum_j y_j(a_{je} - R_e a_{jr}) + \sum_c y_c(a_{ce} - R_e a_{cr}) \quad (k = 1, \ldots, M-1),$$
$$(4.2.18)$$

and, corresponding to equation (4.2.15):

$$F_k = \sum_j y_j + \sum_c y_c - 1 \quad (k = M). \qquad (4.2.19)$$

For convenience we combine these equations into the set of equations:

$$F_k = \sum_j a_{jk} y_j + \sum_c a_{ck} y_c. \qquad (4.2.20)$$

In the Brinkley method, the mole fractions of the components c are estimated (y_c) and the mole fractions of the derived species (y_j) are related to the y_c through the equation (4.2.17), written as:

$$y_j = K_j \prod_c y_c^{\nu_{jc}}. \qquad (4.2.21)$$

Considering the y_c as independent variables, expanding the functions F_k as a Taylor series around the estimates y_c, and then truncating after the first-order terms, leads to:

$$0 = F_k + \sum_c \frac{\partial F_k}{\partial y_c} \delta y_c. \qquad (4.2.22)$$

From equations (4.2.20) follows:

$$\frac{\partial F_k}{\partial y_c} = \sum_j a_{jk} \frac{\partial y_j}{\partial y_c} + a_{ck}, \qquad (4.2.23)$$

where $\partial y_j / \partial y_c$ is obtained by differentiation of (4.2.21), viz:

$$\frac{\partial y_j}{\partial y_c} = \frac{\nu_{jc}}{y_c} y_j. \qquad (4.2.24)$$

On substitution of equations (4.2.23) and (4.2.24) into (4.2.22) there is obtained the set of M equations:

$$0 = F_k + \sum_c \left[a_{ck} + \sum_j a_{jk} \frac{\nu_{jc}}{y_c} y_j \right] \delta y_c. \qquad (4.2.25)$$

It is useful to introduce the abbreviation:

$$r_{ck} = a_{ck} y_c + \sum_j a_{jk} \nu_{jc} y_j \qquad (4.2.26)$$

with which equations (4.2.25) become:

$$\sum_c r_{ck} \frac{\delta y_c}{y_c} = -F_k. \qquad (4.2.27)$$

Equations (4.2.27) form a set of M linear equations in the corrections δy_c, from which the δy_c can be solved by a variety of standard methods. The preferred method for the computer solution of these equations is the well-known Crout reduction method (Crout, 1941).

The corrections δy_c to the estimates y_c then lead to new estimates of the solutions x_c:

$$(y_c)_{\text{new}} = y_c + \delta y_c. \tag{4.2.28}$$

As was mentioned earlier, the corrections may be so largely negative that negative mole fractions are calculated with (4.2.28). Equation (4.2.27) shows clearly how the introduction of logarithmic variables overcomes this problem. Since the Newton–Raphson technique is a first-order approximation, one can use as approximation for $\delta y_c / y_c$ the function $\delta \ln y_c$, which is equal to $\delta y_c / y_c$ to first order. Hence:

$$\sum_c r_{ck} \delta \ln y_c = -F_k \tag{4.2.29}$$

which leads to corrections $\delta \ln y_c$ to the estimates $\ln y_c$, and one obtains for the next set of estimates:

$$(\ln y_c)_{\text{new}} = \ln y_c + \delta \ln y_c$$

or
$$(y_c)_{\text{new}} = y_c \exp (\delta \ln y_c) \tag{4.2.30}$$

and equation (4.2.30) shows clearly that $(y_c)_{\text{new}}$ cannot possibly become $\leqslant 0$. The mole fractions of the derived species then also always remain positive, see equation (4.2.21).

The Newton–Raphson method is a rapidly converging one when it does converge: the errors are to second order. However, the iteration can easily diverge, or there may not even be a unique solution (Hildebrand, 1956). The criterion for divergence is provided by the (Jacobian) determinant of the functions F_k, whose coefficients are the r_{ck} defined by (4.2.26). When the determinant vanishes entirely or almost entirely, the matrix becomes singular or near-singular, and divergence must be expected. For a more detailed analysis of the convergence properties of the Newton–Raphson method, see §5.2.

4.2.3 Inclusion of condensed phases

Brinkley's original treatment was a very general one in that it included consideration of heterogeneous systems. The equations presented in §§4.2.1 and 4.2.2 were all derived for homogeneous, gas phase systems. In the case of the assumed presence of condensed phases the procedure is basically identical, in that equations (4.2.14)–(4.2.17) are all applicable in each phase ϕ. Kandiner and Brinkley (1950) note that each solid species has to be selected as one of the components. The maximum number of condensed phases that can be present, in the case where a gas phase is present, is equal to $M-1$ (cf. §1.3.3), in which case all solids and one gaseous constituent are to be selected as components, and the problem is very much simplified. For the general case of N species, all present in Φ phases, Brinkley (1947) introduced representative phases, one for each species (components and derived constituents), and the concentration of the given species in its representative phase could be used for the computation via the appropriate distribution equilibrium constants, of the concentration of that same species in all other phases. However, the main problem in dealing with condensed phases is that an initial assumption has to be made as to which phases are present. The correctness of this assumption is determined only after an iterative computation has converged, and if wrong, another assumption must be made and the computation must be repeated. Kandiner and Brinkley (1950) illustrate this trial and error approach for the often encountered problem where solid carbon may occur in the equilibrium composition, e.g. of combustion products.

Boll (1961) has extended Brinkley's procedure and devised a method which obviates the necessity for this trial and error procedure. The essential departure from Brinkley's method, proposed by Boll, is that the equilibrium conditions are written as inequalities rather than as equalities. Thus, instead of equation (4.2.15), Boll writes:

$$x^\phi = \sum_j x_j^\phi + \sum_c x_c^\phi \qquad (4.2.31)$$

and uses the inequality: $\quad 1 - x^\phi \geqslant 0 \qquad (4.2.32)$

as the equilibrium condition for phase ϕ. The inequality can be interpreted as follows: If the phase ϕ is correctly assumed to be absent, the equilibrium composition will lead to $1 - x^{\phi} \geqslant 0$, whereas if the phase ϕ is incorrectly assumed to be absent, one obtains $1 - x^{\phi} < 0$ (or $x^{\phi} > 1$).

The problem is set up with the initial assumption that all Φ phases are present. Then, at the rth iteration step, if the value of x^{ϕ} at the end of that step is < 1 for any phase ϕ, that phase is eliminated for the $(r + 1)$th iteration. Conversely, if x^{ϕ} at the end of step r is $\geqslant 1$, the phase ϕ is assumed to be present for the $(r + 1)$th iteration, and the treatment proceeds as in the Brinkley method. The final composition will include the correct phases when the procedure converges.

The treatment described in the above paragraph is suitable also for the consideration of pure condensed species; for such species $x_{\ell}^{\phi} = x^{\phi}$, and x_{ℓ}^{ϕ} should become equal to 1.

4.2.4 Non-ideality considerations

For systems in which the deviation from ideal behaviour is too large to be ignored, equation (4.2.4) is no longer valid. Instead, the analogous equation

$$K_{pj} = \frac{f_j}{\prod\limits_{c} (f_c)^{\nu_{jc}}} \qquad (4.2.33)$$

has to be used. In this expression, the f_j and f_c are partial fugacities, as defined in equation (1.4.15). By using the definition of the activity coefficients (see equation 1.4.18), equation (4.2.33) can be rewritten in a form analogous to that of equation (4.2.17), viz:

$$x_j = K_{pj} \cdot p^{(\sum_c \nu_{jc} - 1)} b_j \cdot \prod_c x_c^{\nu_{jc}}, \qquad (4.2.34)$$

where b_j is an abbreviation for the non-ideality factor:

$$b_j = \frac{\prod\limits_{c} \gamma_c^{\nu_{jc}}}{\gamma_j}, \qquad (4.2.35)$$

where the γ_j, γ_c are activity coefficients. The value of b_j can be established from experimental data of partial molar volumes (see equation 1.4.16), or by an analytical equation of state (see e.g. equation 1.4.5).

For ideal systems, b_j is equal to one. For non-ideal systems the values of the b_j will be functions of temperature, pressure and composition. Thus, the b_j are not known *a priori*. Kandiner and Brinkley (1950) suggest a procedure wherein all the b_j are initially set equal to 1. Having solved the working equations as outlined in §§4.2.1 and 4.2.2, the b_j can then be evaluated from a next best estimated set of component concentrations, for the given specified set of thermodynamic variables (such as p, T). By thus incorporating a recalculation of the b_j at each step in the equilibrium iteration procedure, the process may converge to the final required solution. Usually, the b_j are relatively insensitive to the composition, and, consequently, it is frequently sufficiently accurate to employ the rth approximation to the composition in calculating the b_j to be employed in the $(r+1)$th iteration (Brinkley, 1956).

However, in some instances, the necessity for inclusion of non-ideality in the calculations may lead to a divergent iteration. In such cases it is sometimes possible to find an analytical expression for the b_j in terms of, say, p, T, and the x_j, x_c, and to include the b_j functions in the iterative procedure. This can be illustrated as follows: Assume that use can be made of Lewis' rule (equation 1.4.17), and that the truncated virial equation of state (1.4.6) can be used. Then,

$$b_j = (f/p)^{(\sum_c \nu_{jc} - 1)} \tag{4.2.36}$$

and since

$$\ln \frac{f}{p} = \frac{Bp}{RT} \tag{1.4.14}$$

equation (4.2.34) becomes, by analogy with equation (4.2.21):

$$y_j = K_j . \exp \left[\frac{Bp}{RT} (\textstyle\sum_c \nu_{jc} - 1) \right] \prod_c y_j^{\nu_{jc}}. \tag{4.2.37}$$

The Newton–Raphson procedure, applied to the set of equations (4.2.20) and (4.2.37) then leads to equations similar to equations (4.2.23), but with

$$\frac{\partial y_j}{\partial y_c} = \frac{\nu_{jc}}{y_c} y_j + \frac{p}{RT} (\textstyle\sum_c \nu_{jc} - 1) \frac{\partial B}{\partial y_c} \tag{4.2.38}$$

and the subsequent modification in the solution is only dependent on finding an analytical expression for the function B. Such an expression is given in the form of equation (1.4.7).

4.3   The NASA method

4.3.1   Principle of the method

This method was developed by what is now known as the National Aeronautics and Space Administration (NASA). The first report, describing the method in the form in which it is usually applied, is a publication by Huff *et al.* (1951). The NASA method differs from the Brinkley method in the following important respects: (*a*) the gaseous atoms are arbitrarily selected as components, (thus: condensed species are not, as in the Brinkley method, considered to be components), (*b*) corrections are applied to all constituents, and (*c*) the thermodynamic state of the system can be specified by assigning any two thermodynamic variables selected from: pressure, volume, temperature, enthalpy, entropy, etc. The latter point may appear to be trivial, but it will be seen below that it is a very useful one in that any one of these thermodynamic variables can be considered as an iteration variable in exactly the same sense as the composition variables.

The treatment of the NASA method described below is the one based on the method of Huff *et al.* (1951), as modified by later NASA investigators (Zeleznik and Gordon, 1960, 1962).

Gaseous atoms are selected as components. This means that all derived species are either molecular compounds or condensed species. All these derived species present in the mixture can be represented by the formation reaction given by equation (4.2.3), with $c = e$:

$$\sum_e \nu_{je} \mathscr{B}_e = \mathscr{A}_j \qquad (4.3.1)$$

for which the mass action expression (4.2.4) holds. Note that in this case, the stoichiometric coefficients ν_{je} are the same as the atomic coefficients a_{je}. In logarithmic form, (4.2.4) becomes:

$$\ln K_{pj} = \ln p_j - \sum_e \nu_{je} \ln p_e, \qquad (4.3.2)$$

where p_j and p_e are the partial pressures of the derived species j and of gaseous elements (components) e respectively. For pure condensed species $p_j = 1$ and $\ln p_j = 0$.

For estimated compositions, the right-hand side will generally not be equal to $\ln K_{pj}$ but to, say, $\ln Q_j$, and one obtains for the discrepancies:

$$\Delta g_j = \ln Q_j - \ln K_{pj} = \ln p_j - \sum_e \nu_{je} \ln p_e - \ln K_{pj}. \quad (4.3.3)$$

The thermodynamic significance of Δg_j is given by the expression:

$$\Delta g_j = \frac{(\Delta G_T)_j}{RT}, \quad (4.3.4)$$

where $(\Delta G_T)_j$ is the Gibbs free energy change across the reaction j (4.3.1); when equilibrium is reached, all $(\Delta G_T)_j$ become o. It is shown in chapter 1 that $\ln K_{pj} = (\Delta G_T{}^\circ)_j/RT$ (equation 1.2.33). The atom balance equations (4.2.4) can now be written as:

$$n_e + \sum_j \nu_{je} n_j = B_e, \quad (4.3.5)$$

where, as was seen above: $\nu_{je} = a_{je}$. Huff *et al.* also replace B_e by $a_e A$, where a_e are the number of gram atoms of element e per equivalent formula of reactant, and A is the number of formula weights of reactant. Although this is an unessential variation of the method, we will include it in the following, in order to facilitate comparison with the references dealing with the NASA method. Thus:

$$a_e = \frac{n_e}{A} + \frac{1}{A} \sum_j \nu_{je} n_j. \quad (4.3.6)$$

For an estimated set of n_i values, the right-hand side will usually not be equal to a_e but to, say $a_e + \Delta a_e$, thus:

$$\Delta a_e = \frac{n_e}{A} + \frac{1}{A} \sum_j \nu_{je} n_j - a_e. \quad (4.3.7)$$

In addition, assume that the total pressure is specified (p) and that estimates of the n_i correspond with a pressure $p + \Delta p$. Then:

$$\Delta p = \sum_g p_g - p. \quad (4.3.8)$$

Also assume that, for example, the enthalpy is specified (H), as in the case of an adiabatic process (see §1.6.2), and that the estimates correspond with an enthalpy $H + \Delta H$; then:

$$\Delta H = \frac{1}{A} \sum_i n_i (H_T{}^\circ)_i - H. \quad (4.3.9)$$

(Alternative sets of specified thermodynamic variables are treated in §4.3.2).

Huff *et al.* also assume that the reaction products are contained in a volume $V = RT$, such that, for ideal gases:

$$p_g = n_g. \tag{4.3.10}$$

With this assumption, the set of expressions (4.3.3), (4.3.7), (4.3.8) and (4.3.9) can be considered to be a set of

$$(N-M)+M+1+1 = N+2$$

non-linear equations in the $N+2$ variables n_j, n_e, A and T. (The $N-M$ variables n_j are those of condensed species s (S in number), and of gaseous molecular compounds m ($N-M-S$ in number).) The Newton–Raphson method (see §4.2.2) for solving this set consists in expanding the functions Δg_j, Δa_e, Δp and ΔH in a Taylor series about an estimated set of values of the variables, and truncating after the first-order terms. The following set of simultaneous linear correction equations then results:

$$\Delta \ln n_m - \sum_e \nu_{me}\Delta \ln n_e - q_m \Delta \ln T = -\Delta g_m \tag{4.3.11}$$

($N-M-S$ equations for gaseous molecular compounds m)

$$-\sum_e \nu_{se}\Delta \ln n_e - q_s \Delta \ln T = -\Delta g_s \tag{4.3.12}$$

(S equations for condensed species)

$$\sum_m \nu_{me}n_m\Delta \ln n_m + n_e\Delta \ln n_e + \sum_s \nu_{se}\Delta n_s - Aa_e\Delta \ln A = -A\Delta a_e \tag{4.3.13}$$

(M equations for the elemental abundances)

$$\sum_m n_m\Delta \ln n_m + \sum_e n_e\Delta \ln n_e = -\Delta p, \tag{4.3.14}$$

$$\sum_m n_m(H_T^\circ)_m\Delta \ln n_m + \sum_e n_e(H_T^\circ)_e\Delta \ln n_e + \sum_s (H_T^\circ)_s\Delta n_s$$

$$-AH\Delta \ln A + (\sum_i n_i(C_p^\circ)_i)T\Delta \ln T = -A\Delta H, \tag{4.3.15}$$

where q_m and q_s are abbreviations for $(\partial \ln K_p/\partial \ln T)_j$, which can be obtained from thermodynamic data. The $(C_p^\circ)_i$ are the differentials $(\partial H^\circ/\partial T)_i$, also available from reference texts (Stull, 1965).

From this set of $N+2$ linear equations in $\Delta \ln n_m$, $\Delta \ln n_e$, Δn_s, $\Delta \ln T$ and $\Delta \ln A$ the corrections to the estimates n_i (i.e. n_m, n_e, n_s), T, A can be calculated. Huff et al. described the method of solution of the $N+2$ order matrix via the Crout auxiliary matrix method (Crout, 1941; Kunz, 1957). However, the matrix notation used is less convenient than the purely algebraic notation for the same method given by Zeleznik and Gordon (1960), which we will use here.

Equations (4.3.11) are explicit expressions for $\Delta \ln n_m$ of the $N-M-S$ gaseous molecular compounds. Substitution of $\Delta \ln n_m$ in equations (4.5.12)–(4.5.15) gives the following set of simultaneous equations in $\Delta \ln n_e$, Δn_s, $\Delta \ln T$, $\Delta \ln A$:

$$\sum_e \nu_{se} \Delta \ln n_e + q_s \Delta \ln T = \Delta g_s, \qquad (4.3.16)$$

$$n_f \Delta \ln n_f + \sum_e r_{ef} \Delta \ln n_e + \sum_s \nu_{se} \Delta n_s - A a_e \Delta \ln A + q_e \Delta \ln T$$
$$= \Delta g_e - A \Delta a_e, \qquad (4.3.17)$$

$$\sum_e \{n_e + \sum_m \nu_{me} n_m\} \Delta \ln n_e + (\sum_m n_m q_m) \Delta \ln T = \sum_m n_m \Delta g_m - \Delta p, \qquad (4.3.18)$$

$$\sum_e \{n_e (H_T^\circ)_e + \sum_m \nu_{me} n_m (H_T^\circ)_m\} \Delta \ln n_e + \sum_s (H_T^\circ)_s \Delta n_s - A H \Delta \ln A$$
$$+ (C_p^\circ + \sum_m n_m (H_T^\circ)_m q_m) \Delta \ln T = \sum_m n_m (H_T^\circ)_m \Delta g_m - A \Delta H, \qquad (4.3.19)$$

where f is an auxiliary subscript, to denote element e, and where the following abbreviations have been introduced:

$$r_{ef} = \sum_m \nu_{me} \nu_{mf} n_m, \qquad (4.3.20)$$

$$q_e = \sum_m \nu_{me} n_m q_m, \qquad (4.3.21)$$

$$\Delta g_e = \sum_m \nu_{me} n_m \Delta g_m, \qquad (4.3.22)$$

and
$$C_p^\circ = \{\sum_i n_i (C_p^\circ)_i\} T. \qquad (4.3.23)$$

Thus, the set of equations (4.3.11)–(4.3.15) has been reduced to a set of $M+S+2$ equations in the $M+S+2$ corrections: $\Delta \ln n_e$, Δn_s, $\Delta \ln A$ and $\Delta \ln T$. This set of equations can be solved by standard matrix inversion methods (Hildebrand, 1957). The corrections lead to new estimates of n_e, n_s, A and T. On back-substitution into equation (4.3.11), corrections $\Delta \ln n_m$ are found, which lead to new estimates of the n_m. When the temperature is assigned instead of the enthalpy, the problem becomes somewhat simpler: equation (4.3.19) is no longer needed and $\Delta \ln T = 0$ in equations (4.3.16)–(4.3.18).

The NASA method has several advantages over the original Brinkley method: (a) it does not require estimation of a set of components, (b) the initial set of estimates need neither obey the mass balance equations nor the mass action equations, (c) solids are taken into account very easily. However, Zeleznik and Gordon (1960, 1968) have made a thorough analytical study of the NASA and Brinkley methods, and have shown that both methods, and particularly the Brinkley method, can be modified to allow a direct comparison. This comparison then shows that the methods are computationally equivalent. It should also be pointed out here that Zeleznik and Gordon included the RAND method (see §3.3.2) in their investigation, suitably modified to allow estimates not obeying the atom balances and to include condensed species. It was shown that the RAND method is also essentially equivalent, computationally, to the NASA method, and hence to the Brinkley method. The comparison made by Zeleznik and Gordon (1960, 1968) is discussed in some more detail in §§5.2, 5.3 and 5.5.

4.3.2   Assignment of thermodynamic variables

Usually, equilibrium computations are performed at assigned pressure and temperature, or, sometimes at assigned volume and temperature. This is the case even when adiabatic reaction products are calculated, e.g. at constant pressure: the temperature is estimated, the chemical equilibrium is computed at p and T, and a better temperature estimate is calculated from the adiabaticity relation. In other words: the temperature iteration is performed by an outside loop in the computational algorithm (cf. §1.6.2).

However, in the NASA method the temperature can be determined directly during the iterative process by an iterative loop that is indistinguishable from the iterative loops for the computation of the equilibrium composition. In section 4.3.1. above this was done for an isenthalpic (constant pressure, adiabatic) process. Very similar procedures can be followed for all other processes of practical interest. For example, in the calculation of the performance of explosives, the explosion properties are calculated by assuming an iso-energetic (constant volume, adiabatic) process. The performance calculations for rocket propellants require the assumption of isentropic expansion through a nozzle. In this case, the assigned thermodynamic variables are, usually: entropy and pressure. The isentropicity condition introduces an equation replacing equation (4.3.9) above by:

$$\Delta S = \frac{1}{A} \sum_i n_i (S_T^\circ)_i - S \qquad (4.3.24)$$

and this leads to the entropy equivalent of equation (4.3.15):

$$\sum_m n_m (S_T^\circ)_m \Delta \ln n_m + \sum_e n_e (S_T^\circ)_e \Delta \ln n_e + \sum_s (S_T^\circ)_s \Delta n_s$$
$$- A S \Delta \ln A + \{\sum_i n_i (C_p^\circ)_i\} \Delta \ln T = -A \Delta S$$
$$(4.3.25)$$

which leads to an equation replacing (4.3.19) in the set of equations (4.3.16)–(4.3.19) to be solved for the corrections.

Huff et al. (1951) give examples of several other combinations of assigned thermodynamic variables; some of these are of considerable interest, e.g. entropy and sound velocity (for isentropic expansion to the velocity of sound); entropy and Mach number, entropy and throat area (for isentropic expansion in a designed rocket nozzle), etc. Brinkley, in later modifications of his own method, considers in a systematic manner the following pairs of assigned thermodynamic variables: (T, ρ); (T, p); (H, p), (S, p), and shows that the Brinkley method is also capable of carrying out iterations on these without resorting to outside loops in the computational algorithm (Brinkley, 1960).

4.3.3 Modifications of the NASA method

Several modifications of the above-described method have been published by investigators other than NASA staff.

Michels and Schneiderman (1963) describe a computer program wherein non-ideal, homogeneous gas systems are treated by a method which is essentially the same as the NASA method, with correction equations (4.3.16)–(4.3.19) modified to take into account gas imperfection terms in the form of virial coefficients. The equation of state used by Michels and Schneiderman is equation (1.4.4) with all terms up to and including D retained. The partial fugacities, f_i, are expressed, as in equation (1.4.16) in terms of partial volumes, which in turn are related to the second, third and fourth virial coefficients. The non-ideality function b_j (see equation 4.2.35) becomes an analytically explicit function in the composition, and in p, T, and the interaction coefficients between the various constituents in the composition. The functions b_j, as well as the non-ideality contributions for the pressure (deviation from Dalton's law), for the enthalpy and for the entropy, are all expanded in a Taylor series, along with the expansion of the basic set of equations for the ideal gases, to yield a corrected set of equations (4.3.16)–(4.3.18) and (4.3.19) or (4.3.25). The interested reader is referred to the paper by Michels and Schneiderman for the details of the equations thus derived.

McMahon and Roback (1960) dealt with problems encountered in temperature regions where phase transitions occur; a given condensed species is considered to be absent at a temperature above its melting (or boiling) point. This can present some problem if, during later stages of the computation, the temperature would be computed to be lower and the condensed species has to be considered again. This is, however, simply a tedious book-keeping problem, which only requires additional memory space in the computer used. Mentz (1960) deals with condensed species by comparing vapour pressure values of the condensing species with input equilibrium vapour pressure data for the species, until the equilibrium point is reached.

The NASA method, being a Newton–Raphson method, suffers from serious convergence problems when the initial estimates are not close to the solution (see §5.2). Various modifications have been proposed. The simplest one is a reduction of the computed corrections by a constant factor (Mentz, 1960). A more sophisticated approach was suggested by Tsao and Wiederhold (1963), who used a variable reduction factor, w_i, to give corrected $\ln x_i$ values as follows:

$$(\ln x_i)_{\text{new}} = (\ln x_i) + w_i \Delta \ln x_i, \qquad (4.3.26)$$

where the $\Delta \ln x_i$ are the computed corrections. Tsao and Wiederhold also put:
$$W = \sum_i w_i \Delta \ln x_i. \qquad (4.3.27)$$

When the correct answer is obtained, W and the $\Delta \ln x_i$ are close to zero. This suggests a procedure whereby, at each step, a check is made to see whether W has been decreased. When W is decreasing, the iteration proceeds smoothly. However, when W is increasing, a subsidiary iterative procedure can be applied that endeavours to minimize W by applying optimum correction factors w_i. Wilkins (1960) used a crude, but effective, equation to obtain better estimates y_j of the mole fractions x_j, viz:

$$(\ln y_c)_{\text{new}} = \ln y_c + \frac{1}{L_c} \ln \frac{B_e}{B'_e}, \qquad (4.3.28)$$

where B_e is the number of gram atoms of element e in the system, B'_e is the number of gram atoms of element e calculated by inserting the estimates y_c in the mass balance equations; L_c is the order of the polynomial in the y_c obtained by inserting the mass action equations in the mass balance equations. Thus, equation (4.3.23) uses the crude approximation that all terms with degree less than L_c are dropped. The new estimates y_c can then be used in the NASA method. Barnhard and Hawkins (1963) proposed a modification wherein, before each division involved in solving the set of equations (4.3.13)–(4.3.16), the divisor is checked; if it is zero, the corresponding $\Delta \ln x_i$ is set equal to zero, and the solution is proceeded with. The $\Delta \ln x_i$ are then obtained by back-substitution.

A more detailed discussion of convergence characteristics, amongst others of the NASA method, is given in §5.2.

4.4 Trial and error methods

In §§4.2 and 4.3 iteration methods for solving the equilibrium problem were discussed. These methods suffer, in many instances, from the disadvantage that the initial estimates have to be sufficiently close to the solution, for convergence to take place. Analytical convergence criteria can be established (see §5.2), but a detailed examination of such criteria requires time and/or computer memory space.

Trial and error methods have been developed, which can be made to converge in all instances. Two of these methods will be discussed in this section.

4.4.1   Component reduction method

A systematic trial and error method was developed by Scully (1962), who used the basic working equations (4.2.4) and (4.2.6) as a starting point. Scully followed the Brinkley method by introducing components (c) and derived species (j), whose mole numbers, n_c and n_j respectively, obey the atom balance conditions (4.2.6) written in the form (4.2.9):

$$n_c = q_c - \sum_j \nu_{jc} n_j. \qquad (4.4.1)$$

It is from this point on that Brinkley's and Scully's methods differ. Brinkley's method is iterative in that n_c is estimated, initially equal to q_c and at subsequent stages from new values for n_c calculated from (4.4.1). However, Scully does not use the n_c values following from (4.4.1) as next estimates, but calls the right-hand side of equation (4.4.1): d_c

$$q_c - \sum_j \nu_{jc} n_j = d_c. \qquad (4.4.2)$$

Instead of using d_c as an estimate, Scully uses an estimating equation which makes use of the knowledge that q_c is the maximum value that n_c can have. The equation reduces q_c by pre-assigned amounts, viz:

$$n_c = q_c - \theta \lambda \qquad (4.4.3)$$

and it is this reduction which led us to give Scully's method the name: component reduction method. In (4.4.3), θ is a positive integral index, and λ is a step length. Equation (4.4.3) is used to estimate n_c at each step; then the mole numbers n_j, of the derived species, are calculated from equation (4.2.10), d_c is calculated from equation (4.4.2), and a comparison is made between d_c and n_c, to see whether the inequality

$$n_c > d_c \qquad (4.4.4)$$

is obeyed. Subsequent steps are taken by increasing the index θ, keeping λ constant, until the inequality (4.4.4) is no longer obeyed. Then, subsequent reductions are made by decreasing λ, and the procedure is continued until, again, the inequality no longer holds, at which stage λ is again decreased, etc. The process of reducing n_c until d_c is no longer less than n_c is continued until the lowest n_c has been obtained which is consistent with (4.4.4). This is normally done until a prescribed tolerance τ is reached and:

$$\left| n_c - d_c \right| < \tau.$$

The above trial and error procedure applies to a given component species c. It is subsequently applied to all other component species, in series, until one pass has been made through all M species c. The set of n_c values thus determined is not yet equal to the equilibrium set, since for each component the procedure involves the mole numbers of all other components. A serial order of adjustments, involving the checking of each of the M inequalities (4.4.4), can be selected to guarantee convergence to the equilibrium concentrations (see below).

Scully mentions that one possible difficulty is that the value of the sum of the numbers of gaseous moles, n, changes whenever another n_c value is estimated; however, this is not usually a problem. Furthermore, even if this would lead to a divergent iteration routine, the problem could easily be overcome by utilizing mole fractions rather than mole numbers, in other words: equation (4.2.16) should be used instead of equation (4.2.9) (or (4.4.2)).

A further difficulty in Scully's method is associated with the step length λ. At each stage, for a given value of λ, θ should be increased until (4.4.4) no longer applies, and only then should λ be decreased. The value of λ is usually decreased by multiplying by a factor of $0 \cdot 1$. The refinement, at subsequent stages, has to be made dependent on the accuracy, or tolerance τ, required for the n_c. The difficulty in the method is that, by the time the last refinement has been made (say $\lambda = 0.0001$ times the original λ value), the value of θ then attained can have built up to prohibitively large numbers, or in other words: convergence may be too slow for convenience. A check procedure involving comparison of q_c and n_c has been devised by Scully to overcome this difficulty. The initial value of q_c must be taken to mean the initial value for a particular step length λ. A comparison of q_c and n_c is made, q_c being modified for

$$q_c - n_c > \lambda. \tag{4.4.5}$$

This can be done by setting all $q_c = n_c$ followed by another pass through the calculations for all c. There may then be formed a new set of n_c, some or all being smaller than the previous n_c obtained, but none being greater. When the comparison is such that (4.4.5) no longer holds, the reduction of n_c is complete at the particular λ level in question, and the procedure can be repeated at the next refined level $\lambda = 0.1\lambda$, where initially the q_c are set equal to the last computed values of n_c.

The main advantage of the component reduction method is that it always converges, although it may be quite slow as compared, for example, to a Newton–Raphson interation in the Brinkley method. The method is particularly slow when some of the derived species are present in amounts comparable to or larger than the amounts of one or some of the components.

The method requires the initial solution of the set of mass balance equations to give the q_c. Drifting cannot take place, because the q_c, once established, retain the same accuracy, and the reduction equation (4.4.3) does not contribute to drift.

Condensed species can be taken into account very simply. As in Brinkley's method, possible condensed species are considered as components. Each condensed species, if included, requires consideration of one additional mass balance equation. Thus, the condensed component reduction proceeds according to equation (4.4.3) until the value of n_c becomes equal to d_c (in which case the condensed species is present and the mass action equation holds) or until n_c becomes $\leqslant 0$ (in which case the condensed species is absent and the mass action equation has to be dropped).

One serious disadvantage of the method is that, when a survey of computations of equilibria, e.g. in various temperature and pressure ranges is required (cf. §2.5), the calculation has to be restarted from q_c in each (p, T) case, in order to allow the reduction of q_c to lead to the equilibrium n_c. In other words: the equilibrium n_c values for one (p, T) problem cannot serve even as an estimate for another (p, T) problem, even if the variables p and/or T in the two problems are very close to each other.

4.4.2 Sliding target method

This method (*Methode des Gleitenden Ziels*) was proposed by Horn and Troltenier (1962) as a modification of the method previously described by Horn and Schüller (1957), which in turn is based on Brinkley's method (§4.2). Horn and Troltenier use partial pressures in their method, but in our description we will discuss the method in terms of mole fractions.

Suppose a set of estimates is made of the mole fractions, y_c, of the components. In the solution via the Brinkley method, this set of estimates may or may not lead to convergence. Suppose that, for the specified elemental abundances, B_e (initial target), the estimates do not result in a convergent iteration. However, the estimates may well lead to convergence for a different target of elemental abundances. Horn and Troltenier (1962) propose the following method for the selection of a new target:

The estimates y_j correspond with:

$$y = \sum_c y_c + \sum_j y_j \qquad (4.4.6)$$

instead of with equation (4.2.15):

$$1 = \sum_c x_c + \sum_j x_j \qquad (4.4.7)$$

and with B'_e instead of with B_e (equation 4.2.6). Then:

$$\left. \begin{aligned} (y)_1 &= \tfrac{1}{2}(1+y), \\ (B_e)_1 &= \tfrac{1}{2}(B_e + B'_e) \end{aligned} \right\} \qquad (4.4.8)$$

are selected as the new target, with which the iteration can be tried again for the same initial estimates y_c. If convergence is

still not attained, a further halving of the interval may be made
to find another new target, viz:

$$\left.\begin{array}{c}(y)_2 = \frac{1}{2}[(y)_1+y], \\ (B_e)_2 = \frac{1}{2}[(B_e)_1 + B'_e].\end{array}\right\} \qquad (4.4.9)$$

This procedure for the selection of a new target can be repeated
until a target is found with which the initial set y_c leads to a
convergent iteration. If, for any new target, convergence is
attained, the solution for that new target can then be used as a
new initial set of estimates y_c with which an attempt can be made
to converge to the initial, specified target again. In this manner,
by finding new targets that come successively closer to the initial,
specified target, in the direction guided by convergence or
divergence, convergence will always take place eventually.

The above described procedure is very similar to the parameter-
perturbation procedure for the solution of sets of non-linear
equations, proposed by Freudenstein and Roth (1963). This
procedure is a numerical adaptation of the implicit function
theorem, wherein a set of 'derived' equations, with one known
root, is deformed into the set of equations one desires to solve.

4.5 The method of successive reaction adjustments

4.5.1 Principle of the method

This method was first proposed by Villars (1959, 1960) who
developed the method at the Naval Ordnance Test Station
(NOTS) in China Lake, California.[*] Villars' method is similar
to Brinkley's in one respect: a number of components (usually
equal to the number of elements) is selected and then all other
species are expressed in terms of the components by using the
reaction equilibria:

$$\sum_c \nu_{jc}\mathscr{A}_c \rightleftharpoons \mathscr{A}_j. \qquad (4.5.1)$$

For these equilibria, equations (4.2.4) apply, written in the form:

$$\frac{n_j n^{\xi j}}{\prod_c n_c^{\nu_{jc}}} = K_{pj}p^{\xi j}, \qquad (4.5.2)$$

---

[*] Since most of the modifications of this method were also developed at NOTS
the method is often called after this institution: the NOTS method.

where ξ_j is an abbreviation for $\sum\limits_c \nu_{jc} - 1$. The right-hand side of (4.5.2) is abbreviated to K_{xj} (equilibrium constant expressed in terms of mole fractions). For systems at constant T and p the K_{xj} are constants, whose values are given in standard thermodynamic data texts (Stull, 1965). At equilibrium, the values of n_j and n_c have to obey equation (4.5.2).

Villars then proceeds as follows: an initial estimate is made of all n_j and n_c, obeying the atom balance equations (4.2.6). These estimates, which we call m_j and m_c, generally do not obey equations (4.5.2), but the left-hand side will be equal to, say, Q_j:

$$\frac{m_j m^{\xi_j}}{\prod\limits_c m_c^{\nu_{jc}}} = Q_j. \qquad (4.5.3)$$

The discrepancies, D_j, between calculated equilibrium constants Q_j and given equilibrium constants K_{xj}, are then given by:

$$D_j = \frac{Q_j}{K_{xj}} - 1. \qquad (4.5.4)$$

(In a later modification of his method, Villars uses $\ln(Q_j/K_{xj})$ instead of D_j (Villars, 1960).) Subsequently, a correction is made in the mole numbers for that reaction j for which D_j is largest. This is done by introducing the reaction parameter Δm_j which corrects the extent of reaction j by making the corrected mole numbers obey the equilibrium relationship (4.5.2):

$$\frac{(m_j + \Delta m_j)\{m - \xi_j \Delta m_j\}^{\xi_j}}{\prod\limits_c (m_c - \nu_{jc}\Delta m_j)^{\nu_{jc}}} = K_{xj}. \qquad (4.5.5)$$

Equation (4.5.5) is a non-linear equation in Δm_j from which Δm_j can be calculated. Then, with Δm_j, all m_c values in the composition are corrected to $m_c - \nu_{jc}\Delta m_j$, and the new m_c values, together with the one corrected m_j value and the other, still uncorrected m_j values form a new set of estimates with which the above-mentioned correction procedure can be repeated, i.e. another reaction j is selected on the basis of the largest discrepancy D_j corresponding to equation (4.5.4), etc. The procedure is continued until the largest discrepancy is less than a prescribed error, τ. Then $m_j = n_j$ and $m_c = n_c$. One important advantage of Villars' method is that, once the initial set of estimates is made to obey the atom balance equations, all subsequent sets of estimates also obey those equations. The disadvantage is, of course, that the initial set has to be made to obey the atom balances.

Another, considerable advantage is that convergence is assured, since each iteration is equivalent to arresting all possible reactions but one and allowing that one to proceed, according to the law of mass action, to its equilibrium. This possible kinetic model can lead only in the direction of chemical equilibrium.

Also, Villars' method does not require explicit matrix inversion, which means that computer memory requirements are small and matrix near-singularity is not a problem.

On the other hand, Villars' method depends greatly on the choice of the components; when components are chosen which are present in small percentages the procedure is quite slow.

Possible condensed species can be dealt with as shown in §2.4, i.e. by considering one fewer mass balance equation of type (2.4.8) and one more mass action equation, for each condensed species to be considered.

The method proposed by Villars relies on the solution of the equation (4.5.5) to give the required values of Δm_j. This solution can be simplified as follows: equations (4.5.3) and (4.5.5) are inserted into equation (4.5.4), giving:

$$D_j = \frac{\prod\limits_c \left(1 - \nu_{jc}\dfrac{\Delta m_j}{m_c}\right)^{\nu_{jc}}}{\left(1 + \dfrac{\Delta m_j}{m_j}\right)\left(1 - \xi_j \dfrac{\Delta m_j}{m}\right)^{\xi_j}} - 1. \qquad (4.5.6)$$

4.5.2 Approximations

Equations (4.5.6), which are the equivalent of equations (4.5.5) can be solved in principle for Δm_j. However, this is, in general, not an easy problem. Villars chose to use the approximation: $(\Delta m_j/m) \cong 0$, which leads to a considerable simplification.

Goldwasser (1959) describes a method which is basically the same as the above method by Villars. However, Goldwasser only considers first-order solutions to equation (4.5.6). In his paper on the subject, Goldwasser only describes his method with an example. We will consider his treatment in a more general way, as follows:

Consider $\Delta m_j \ll m_j$, and since usually $m_j < m_c < m$, we can then, to first-order, put:

$$D_j = \frac{1 - \Delta m_j \dfrac{\sum\limits_{c} v_{jc^2}}{m_c}}{1 + \dfrac{\Delta m_j}{m_j} - \xi_j^2 \dfrac{\Delta m_j}{m}} - 1 \qquad (4.5.7)$$

which is a linear expression in Δm_j from which Δm_j can easily be calculated. The approximation procedure can be repeated when Δm_j turns out to be large compared to m_j, before the selection of another reaction equilibrium on the basis of largest D_j. The only disadvantage of the approximation procedure is that it can cause drifting away from the elemental abundances. Such drifting requires periodic resetting of the estimates as the solution is approached.

4.5.3 The optimized basis of components

An analytical criterion exists for the determination of the number of components (Brinkley, 1946, 1966), see §1.3.2. An analytical criterion has also been derived, by the NOTS staff, for the selection of those components that are most likely to be present in largest concentrations in the equilibrium mixture to be computed. This selection leads to what has been named the optimized basis of components (Browne *et al.* 1960). Cruise (1964) has modified this optimized basis concept slightly, and has used it in conjunction with the Villars method to obtain more rapid convergence in the successive approximations of that method.

In Cruise's modification of the criterion, an estimate is first made of all mole numbers n_i (these are called m_i, as above). The m_i are sorted so that the molar amounts are in descending order. Then, Brinkley's procedure is followed: A matrix is formed of the vector elements a_{ie} corresponding with the formula vector α_i of substance i. From the matrix, determinants can be formed by taking rows, in sequence, starting from row $i = 1$. Only the largest ordered non-zero determinants that can be formed correspond with component selections. The order of such determinants is equal to the rank of the matrix, which is $\leqslant M$, the number of elements in the system. In this sequence, the first such non-zero determinant that can be formed from the matrix corresponds with formula vectors of the optimized basis of components species (c).

4.5.4 The Cruise–Villars method

Cruise has used the optimized basis concept in conjunction with the Villars method (Cruise, 1964) to obtain a method which gives extremely rapid convergence. The optimized basis is re-selected after each new estimate of n_i has been calculated. The derived species are further subdivided into major derived species (which include condensed species) and minor derived species (whose mole numbers are at least two orders of magnitude smaller). For the major derived species j, Cruise uses a further simplification of equation (4.5.7), viz:

$$\Delta m_j = \frac{\ln (K_{xj}/Q_j)}{\sum_c \zeta_j v_{jc}^2/m_c + \zeta_i/m_j}, \qquad (4.5.8)$$

where $\ln (K_{xj}/Q_j)$ is, to first-order accuracy, equal to D_j, and ζ_j is the phase parameter (1 for a gaseous species, 0 for a pure condensed species, see equation (2.4.3)). For the minor derived (gaseous) species j, Cruise considers that an even greater approximation is justified, viz:

$$\Delta m_j = m_j \left(\frac{K_{xj}}{Q_j} - 1 \right). \qquad (4.5.9)$$

Equation (4.5.9) constitutes considerable saving of time as compared to the usually unnecessarily accurate equation (4.5.8).

One important deviation from Villars' method, introduced by Cruise, is that all major derived species are corrected during one iteration, and not only the one for which D_j is largest. In this modified form, the method becomes very similar to a modification of the Brinkley method, introduced by Kaeppeler and Baumann (1957).

Equation (4.5.8) is a very useful one for the consideration of solids. As treated by Cruise, solids are major derived species, whose corrections during one iteration are computed with equation (4.5.8). When the mole number of the condensed species becomes negligibly small, and ln (K_{xj}/Q_j) is negative, no correction is applied, and the equilibrium relationship is assumed to become inapplicable. In this way, one condensed species is deleted, and according to the phase rule, a degree of freedom is becoming available which leads to the availablity of the appropriate atom balance condition to put an additional constraint on the other species.

Smith and Missen (1968) have shown that the Cruise–Villars method does not always converge, probably owing to the fact that if initial estimates are very different from the solution the simultaneous solution of equations (4.5.8) for all major m_j values can lead to next estimates that are even farther removed from the equilibrium solutions. Smith and Missen express the Cruise–Villars equations in terms of Gibbs free energy differences (ΔG), and introduce a convergence forcing parameter λ with which they modify the changes Δm_j as follows:

$$(\Delta m_j)_{\text{new}} = \lambda \Delta m_j, \qquad (4.5.10)$$

where λ is made to depend on whether or not the next estimate, calculated from the Δm_j, is nearer to equilibrium. They evaluate the derivative

$$\frac{dG}{d\lambda} = \sum_i \mu_i \frac{dn_i}{d\lambda}$$

for the new equilibrium (at $\lambda = 1$), and put $\lambda = 1$ only when $dG/d\lambda$ is $\leqslant 0$. When the new $dG/d\lambda > 0$ they put:

$$\lambda = \frac{(dG/d\lambda)_{\lambda=0}}{(dG/d\lambda)_{\lambda=0} - (dG/d\lambda)_{\lambda=1}} \qquad (4.5.11)$$

from which follows a value of $(0 < \lambda \leqslant 1)$ which leads to $(\Delta m_j)_{\text{new}}$ values according to (4.5.10) with which converging solutions were obtained by Smith and Missen for all cases studied, including those where the Cruise–Villars method failed to converge.

4.6 The method of nested iterations

A method developed especially for the calculation of a large number of similar equilibria, and for use on a small computer, is the method of nested iterations (Storey and van Zeggeren, 1967).

9

The method is also of general usefulness for small ($<$ 10) sets of equations, and for this reason it is included here.

The method is based on the treatment of a many-variable set of simultaneous non-linear equations as a nested set of single-variable iterations. This treatment then allows the use of recent powerful methods for forcing convergence in single-variable iterations.

As in Brinkley's method, the basic set of equations to be solved is the series of equations, represented by the atom balances (4.2.4), and by the mass action equations written in the form (4.5.2). In the general case where there are N possible species, consisting of M different elements, and where there are S possible (pure) condensed species or solids, the number of equations (4.2.4), represented by equations (2.4.8), is $M - S$, whereas the number of independent equations (4.5.2) is $N - M$, i.e. a total of $(N - M) + (M - S) = N - S$ equations. The unknowns are $N - S$ in number if, for the moment, only the gaseous species are considered (cf. §2.4).

The method of nested iterations consists in rewriting and combining equations (2.4.8) and (4.5.2), to form the functions F_h, which correspond with discrepancies in the mass balance equations (see §2.4):

$$F_h = Q_h - \sum_g b_{gh} K_g \prod_{h'} (n_{h'})^{\nu_{gh'}}, \qquad (4.6.1)$$

where h' is a dummy subscript for h ($h, h' = 1, ..., M - S$), and where the Q_h are linear functions of the elemental abundances, formed by eliminating the condensed species from the atom balance equations.

Thus, the problem of calculating equilibrium compositions has been reduced to the problem of finding that set of values of $n_{h'}$ ($h' = 1, 2, ..., M - S$) which reduces the set of functions F_h simultaneously to zero. The functions F_h will be called residuals in the following description of the method of solution by nested iteration. In this method each of the F_h is associated with one and only one $n_{h'}$. Individual iterations consist of the reduction of an F_h to zero by changing only its associate $n_{h'}$. This is done in a systematic cyclic manner which reduces a greater and greater number of residuals F_h to zero simultaneosly until the solution is obtained. This occurs when a set of $n_{h'}$ has been found which makes all the F_h vanish simultaneously. The procedure is equivalent to equating the indices h' to corresponding indices h. The residuals and their associated variables are then visualized as being put on $(M - S)$ different levels, as seen in table $(4.6.1)$.

TABLE 4.6.1

Level No.	Residual	Associated variable
$M-S$	F_{M-S}	n_{M-S}
$M-S-1$	F_{M-S-1}	n_{M-S-1}
.	.	.
.	.	.
.	.	.
h	F_h	$n_{(h=h')}$
.	.	.
.	.	.
.	.	.
2	F_2	n_2
.	.	.
.	.	.
.	.	.
1	F_1	n_1

The association of variables and residuals is best done, for the purpose of convergence, in such a manner that the expected largest variable n_h is put on the lowest level on which the residual function contains this largest variable. However, the association is not critical, i.e. if it is not known *a priori* what the largest variables are, any initial set of associated variables and residuals will lead to the equilibrium composition, albeit at a much slower rate of convergence than could be obtained for an optimized set of associated variables and residuals.

In the case of a single equation in a single variable (no. 1) a number of iteration procedures are available which make a sequence of guesses at $n_{h=1}$ prescribed by some technique such as a binary search or Newton's method, until a value of $n_{h=1}$ is found which makes the residual F_1 vanish (in practice it is normal to require instead that the absolute value of the residual become less than some acceptable small value τ). To use the same approach on a general set of $M-S$ equations we proceed as follows: First an initial set of values of $n_1, n_2, ..., n_{M-S}$ are estimated. A succession of values of n_1 are found by some convenient single-variable iteration method (Traub, 1964) until F_1 vanishes. Then a second guess is made at n_2, the variable on level 2, using again a single-variable iteration which seeks to make F_2 vanish. If this new value of n_2 does not yet make F_2 zero, the computer returns to level 1 and, using the new value of n_2, repeats the computation on level 1, i.e. finds an n_1 which makes F_1 vanish corresponding to the new n_2. Having done this it goes up to level 2 again and makes a further attempt at finding n_2. Thus, in principle the computer continues to find increasingly better n_2 values on level 2, dropping down to level 1 after each new n_2 to find corresponding n_1 values, until a set (n_1, n_2) is found which makes both F_1 and F_2 vanish. A new value for n_3 on level 3 is then found, attempting to make F_3 vanish. Again, if this n_3 is not successful, the computer returns to level 1 and finds once more a set (n_1, n_2) which makes F_1 and F_2 vanish for the new n_3. The computer thus continues to move on to higher and higher levels until level $M-S$ has been completed, at which stage (since all lower levels are solved at each iteration for n_{M-S}) the solution, i.e. that set of $n_1, n_2, ..., n_{M-S}$ for which all $F_1, F_2, ..., F_{M-S}$ vanish, will have been found.

It should be noted that this technique changes a many-variable iteration problem into a set of single-variable problems, i.e. it treats the iterations as a nested set of iterations, each of which iterates on one variable only. This allows the use of single-variable iteration formulae at all levels, which means that convergence can be forced quite easily (see, e.g. Traub, 1964).

The number of functions and associated variables is equal to $M-S$ only for the case where, given the proportions of the element, the conditions are that volume V and temperature T are fixed, at least when non-ideality of the gas phase is not considered. In the case where p and T are fixed (e.g. Kandiner and Brinkley, 1950) the condition that p be constant requires another equation, viz. (for ideal gases) that the sum of the partial pressures be equal to the total pressure (Dalton's law, see equation 1.2.6). This makes it possible to select one additional level where the residual function can then be written (using $p_g = (n_g/n)p$):

$$F_h = \frac{1}{n}(\sum_g n_g) - 1. \qquad (4.6.2)$$

However, as in the case of Brinkley's method, another approach at constant p is to select mole fractions rather than mole numbers as variables, and to relate the atom balance equations to each other as in equation (4.2.13). This reduces the number of equations to be solved to $M - S$ again, but in many cases it is preferable, for the sake of simplicity in setting up the algorithm, to solve the set of $M - S + 1$ equations, including (4.6.2).

The cases of constant T and V or p, discussed in the previous paragraph, are of limited applicability. A more general case is one in which it is required to compute equilibria resulting from adiabatic reactions at constant volume or pressure; for the calculation of such equilibria it is necessary (see §1.6.2) to introduce the additional function F_h requiring that either energy (for constant V) or enthalpy (for constant p) remain constant. The associated variable in both cases is T. The adiabaticity condition leads to the additional residual equation (see §1.6.2):

$$F_h = \int_{T_0}^{T_q} C_v \, dT - Q_v \qquad (4.6.3)$$

or $$F_h = \int_{T_0}^{T_q} C_p \, dT - Q_p \quad \text{respectively,} \qquad (4.6.4)$$

where C_v and C_p are the total heat capacities of the gas mixtures at constant volume and pressure, respectively, and Q_v and Q_p are the heats of reaction at constant volume and pressure, respectively. T_0 is the initial temperature of the reactants, and T_q is the equilibrium temperature to which the computation should converge.

The method described above is a general purpose method for solving sets of non-linear equations, although for large sets it would be fairly slow. However, the method can be very useful in applications to chemical equilibrium computations, in which the number of equations to be solved is rarely more than about six. The method would be faster than, e.g. a multivariable Newton method, mainly because second or higher order iteration formulae can be used. The method does not require explicit matrix inversion; this means that computer programs based on the method require little computer memory space as compared to other, standard methods for solving non-linear equations.

One possible difficulty that might be encountered in using the method of nested iterations is that if a poor initial guess is made at any level $h > 1$ it may be found that on another, lower level, the iteration has no real root. In general this type of difficulty can be overcome by re-enumerating the levels, and the program can be arranged in such a manner that this can be done without rewriting the program. In cases in which no root can be found, even after re-enumeration, the program can be written such that the value of n_h corresponding to a minimum of $F_{h=h'}$ is found and used for subsequent iterations.

One of the main advantages of the method is that the estimated values of the initial set of n_h need not obey the atom balance equations.

The above treatment shows that pure condensed species can be included very simply in the method. Since it is not known before the computation commences, whether a given solid species is going to be present or not, a decision about inclusion of that solid species has to be arrived at during computation. The most convenient way of doing this is to assume that the solid species is present in the equilibrium composition, and then to test that assumption by means of the particular atom balance equation that was eliminated (see §2.4) in the computations because of the assumption.

4.7 Miscellaneous methods

In this section brief descriptions will be given of a number of methods which are difficult to fit into any of the categories of the previous sections. These miscellaneous methods are all methods suitable for calculations of simple systems ($M \leqslant 4$, say) on desk calculators; they are therefore of considerable use when a computer is not available, or when the turn-around time is too large. The methods are simple to use, but suffer from the disadvantage that, in particular for the more complex systems, convergence can

often not be forced, unless the user has considerable prior knowledge of the system or can put his intuition to good advantage.
There are two categories of miscellaneous methods.

4.7.1 Brute force methods

In brute force methods (the term was coined by Bahn, 1960),
estimates are made of the amounts of principal species c, equal in
number to the number of components of the system, M. The
amounts of the derived species j are calculated with the aid of the
appropriate mass action equations. Each principal species is
associated with one and only one atom balance equation (4.2.6);
thus c is put equal to e, as was done, after due allowance for
condensed species, in the nested iterations method (§4.6). The
set of atom balance conditions (4.2.6) is then tested, to see whether
the estimates do indeed lead to the values of B_e corresponding to
those conditions. This will, in general, not be the case, but the
estimated values will correspond to, say, a set of B'_e values. The
difference between the B'_e and the B_e is then used to find corrected
values of the amounts of the principal species.

Bahn (1960), expressing all equations in terms of mole fractions,
calculated values of $R'_e = B'_e/B'_r$, where r refers to a selected
element. From the R'_e he obtained:

$$(y_c)_{\text{new}} = \frac{y_j}{y} \cdot \frac{R_e}{R'_e}, \qquad (4.7.1)$$

where
$$y = \sum_j y_j + \sum_c y_c. \qquad (4.7.2)$$

Hilsenrath *et al.* (1959), expressing their equations in terms of
mole numbers, found new values of m_c as follows:

$$(m_c)_{\text{new}} = m_c \frac{B_e}{B'_e}. \qquad (4.7.3)$$

In addition, a convergence factor was introduced:

$$(m_c)_{\text{new}} = \lambda_c (m_c)_{\text{new}} + (1 - \lambda_c) m_c, \qquad (4.7.4)$$

where λ_c could assume any value between 0 and 1, depending on
the convergence at each step in the iteration.

Clasen (1965), also using mole numbers, put:

$$\Delta B_e = B_e - \sum_i a_{ie} m_i = \sum_i a_{ie} \Delta m_i, \qquad (4.7.5)$$

where the Δm_i values must be chosen so that $m_i + \Delta m_i \geqslant 0$ for all i. This cannot be guaranteed, but an attempt can be made to choose small values for Δm_i. One way to do this is to minimize the function $\frac{1}{2} \sum_i w_i (\Delta m_i)^2$ subject to equation (4.7.5), where w_i is the relative importance or weighting factor of minimizing Δm_i. This reduces to the problem of finding Lagrangian multipliers χ_e, such that the Lagrangian function becomes a minimum. This leads to:

$$w_i \Delta m_i = \sum_e a_{ie} \chi_e. \qquad (4.7.6)$$

Substitution into (4.7.5) then gives

$$\Delta B_e = \sum_f \left[\chi_f \sum_i \frac{a_{if} a_{ie}}{w_i} \right] \qquad (4.7.7)$$

which can be solved for the χ_f, leading to the corrections Δm_i. Clasen found that, with this 'projection' method, satisfactory results could be obtained by using $w_i = 1/m_i$. The choice of the weighting factor depends to some extent on the available computers, but using the weighting factors above, the corrections become quite simple. The $n_i = m_i + \Delta m_i$ values obtained need not all be strictly positive; if any n_i becomes negative, the method has failed.

Wilkins (1960), using partial pressures in the equations, derived the approximate correction equation:

$$\ln (p_c)_{\text{new}} = \ln p_c + L_c \ln \frac{B_e}{B'_e}, \qquad (4.7.8)$$

where L_c is the highest exponent of p_c in the corresponding mass action equation (cf. §4.3.3).

Potter and Vanderkulk (1960) expressed their equations in terms of both mole fractions and mole numbers. Then, from each difference, $\Delta B_e = B_e - B'_e$, a new value of the associated y_c was evaluated from

$$\Delta \ln y_c = \frac{y-1}{\dfrac{\Delta B_e}{n}}. \qquad (4.7.9)$$

In the last two of the above methods, the new estimates cannot be negative, and therefore these methods are to be preferred. All the above described methods are algebraically very similar to the Brinkley and NASA methods. They differ from those methods, however, in that the computational procedure is much simpler. Also, the estimates can be very crude; after only a very few iterations it will become apparent whether another set has to be selected. In none of the above cases is matrix inversion required.

4.7.2 Reduction to one or two equations

In simple systems ($M \leqslant 5$, say) it is frequently possible to predict which species will be most abundant. Suppose that a number M of such species can be selected as components. These M principal species, again denoted by c, can then serve as the variables in which the M atom balance equations can be expressed, because the remaining $N-M$ derived species j can be expressed in terms of those variables, through the appropriate mass action equations. Thus, equations (4.2.10) can be inserted into equations (4.2.6) to give:

$$B_e = \sum_c a_{ce} n_c + \sum_j a_{je} K_{pj} \left(\frac{p}{n}\right)^{\xi_j} \prod_c n_c^{\nu_{jc}}. \qquad (4.7.10)$$

In all methods described so far, this M-variable set of equations was solved. However, it is often possible to effect a simplification by eliminating one or more variables, particularly for simple sets of equations. Many authors have attempted to derive simpler sets of equations. Under special conditions, and in particular for hydrocarbon fuel/air combustion systems, i.e. for systems consisting of the elements carbon, hydrogen, oxygen and nitrogen, such simple sets of equations have been derived.

We consider it beyond the scope of this book to give derivations of such methods of reduction. It should suffice here to make reference to the various methods that have been successfully applied to combustion equilibria.

A number of articles have appeared in which the reduction to two equations in two unknowns has been shown to be possible. Damköhler and Edse (1943) were the first to derive a successful procedure, which utilized a graphical interpolation method to lead to a convergent iteration (cf. §1.1). Subsequent algebraic variations of the Damkohler and Edse method were published by Sachsel *et al.* (1949), Donegan and Farber (1956) and Harker (1967).

In some cases reduction to a single variable equation is possible. Von Stein (1943) has devised a schematic procedure in which the equations and variables are arranged in tabular form. Guide-lines are given for the elimination procedure which leads to a polynomial equation in one selected unknown variable. Despite considerable effort, von Stein (1943) and von Stein and Voetter (1953) were not able to find simple laws which could be followed for all systems. Each system has to be considered separately, and the user has to acquire considerable skill in order to be able to deduce the final equation. Furthermore, the labour involved in the solution of such a high order single equation is very often prohibitive as well. Similar procedures were followed in initial attempts to compute combustion equilibria, by the Naval Ordnance Test Station (McEwan, 1950), but these methods were soon discarded in favour of the method of successive reaction adjustments, developed at NOTS (§4.5).

4.8   Methods using element potentials

The name 'element potential' was coined by Powell and Sarner, in 1959. Element potentials are used by these authors to derive generalized equilibrium relationships and lead ultimately to a simplified procedure for expressing such relationships. Since element potentials are either explicitly used or explicitly mentioned in several methods of equilibrium computation, a short description of Powell and Sarner's generalized procedure is given here.

At a fixed pressure p and temperature T, the general thermodynamic equilibrium relationship (1.2.23) becomes:

$$dG = \sum_i \mu_i dn_i. \qquad (4.8.1)$$

If the expression (2.3.3), giving μ_i in terms of the undetermined multipliers χ_e, is inserted into (4.8.1), one obtains:

$$dG = \sum_e \chi_e [\sum_i a_{ie} dn_i]. \qquad (4.8.2)$$

From equation (2.2.4) it can be seen that the sum in square brackets is equal to the infinitesimal change in the elemental abundance B_e, thus:

$$dG = \sum_e \chi_e \, dB_e. \tag{4.8.3}$$

Therefore, if the elemental abundances are changed independently, the χ_e can be evaluated as:

$$\chi_e = \left(\frac{\partial G}{\partial B_e}\right)_{p,\,T,\,B_f}. \tag{4.8.4}$$

The element potential is thus defined by Powell and Sarner, in complete analogy with the definition of the chemical potential (cf. §1.2.5), as:

$$\mu_e = \left(\frac{\partial G}{\partial B_e}\right)_{p,\,T,\,B} \tag{4.8.5}$$

so that, from equations (4.8.4) and (4.8.5):

$$\chi_e = \mu_e. \tag{4.8.6}$$

Equation (4.8.6) is valid for all elements $e(1, \ldots, M)$.

Some of the important properties of the element potential can be illustrated as follows: insertion of equations (4.8.6) into equation (2.3.3) gives:

$$\mu_i = \sum_e a_{ie} \mu_e. \tag{4.8.7}$$

Also, substitution of the μ_i from equations (4.8.7) into the expression for the Gibbs free energy (2.2.2), leads to:

$$G = \sum_e \mu_e (\sum_i a_{ie} n_i), \tag{4.8.8}$$

$$G = \sum_e \mu_e B_e. \tag{4.8.9}$$

Equations (4.8.4)–(4.8.9) serve to show the complete analogy between inert and reactive systems, in that all these equations become the general thermodynamic equations for inert systems by replacing B_e by n_i, and the subscript e by i.

Powell and Sarner (1959) point out that μ_e is a function which is characteristic of an element e, throughout a system, regardless of the number and variety of species of which it forms part, of the degree of ionization of those species, and of the distribution over different phases in the system. This is an important conclusion, which has also been derived, via a different route, by Nikolskii (1966). Explicit use of this conclusion has been made by White (1967).

White uses equation (4.8.7) and, by expressing μ_i in the form:

$$\mu_i = \mu_i^\circ + RT \ln p + RT \ln x_i \qquad (1.2.26)$$

obtains
$$\frac{n_i}{n} = x_i = \exp\left[-\frac{\mu_i^\circ}{RT} - \ln p + \frac{1}{RT} \sum_e a_{ie}\mu_e \right] \qquad (4.8.10)$$

or, with the abbreviation (3.2.12):

$$\frac{n_i}{n} = x_i = \exp\left[-c_i + \frac{1}{RT} \sum_e a_{ie}\mu_e \right], \qquad (4.8.11)$$

where μ_e is the element potential of element e or as White calls it: the Gibbs free energy contribution per gram atom of element e.

For a given set of estimates of the μ_e (denoted by μ_e'), the equations (4.8.10) become:

$$\frac{m_i}{m} = y_i = \exp\left[-c_i + \frac{1}{RT} \sum_e a_{ie}\mu_e' \right] \qquad (4.8.12)$$

with the m_i, m and y_i not generally satisfying the atom balance equations and Dalton's law. Thus:

$$\sum_i a_{ie}m_i = B_e' = B_e + \Delta B_e \qquad (4.8.13)$$

and
$$\sum_i y_i = y = 1 + \Delta y. \qquad (4.8.14)$$

In order to convert equation (4.8.13) from mole numbers to mole fractions, White expresses the B_e in terms of ratios to the sum of the B_e, thus:

$$R_e = \frac{B_e}{\sum_e B_e}, \qquad (4.8.15)$$

and hence
$$\Delta R_e = \frac{\sum_i a_{ie}m_i}{\sum_e \sum_i a_{ie}m_i} = \frac{\sum_i a_{ie}y_i}{\sum_e \sum_i a_{ie}y_i}. \qquad (4.8.16)$$

White then uses the standard Newton–Raphson method, i.e. he

expands ΔR_e and Δy in a Taylor series and truncates after the first-order terms. Then, since:

$$\frac{\partial \Delta y}{\partial \mu_f} = \frac{1}{RT} \sum_i a_{if}\, y_i \qquad (4.8.17)$$

and
$$\frac{\partial \Delta R_e}{\partial \mu_f} = \frac{R_e}{RT}\left[\frac{\sum_i a_{ie}a_{if}y_i}{\sum_i a_{ie}y_i} - \frac{\sum_i [\sum_e a_{ie}]a_{if}y_i}{\sum_e B_e}\right] \qquad (4.8.18)$$

the set of correction equations becomes:

$$\sum_f \left\{\sum_i a_{if}\, y_i\right\}\Delta\mu_f = -RT\Delta y \qquad (4.8.19)$$

and
$$\sum_f \left\{\frac{r_{ef}}{\sum_i a_{ie}y_i} - \frac{\sum_e r_{ef}}{\sum_e B_e}\right\}\Delta\mu_f = -RT\frac{\Delta R_e}{R_e}, \qquad (4.8.20)$$

where
$$r_{ef} = \sum_i a_{ie}\, a_{if}\, y_i. \qquad (4.8.21)$$

This is really a set of $M+1$ equations in M unknowns $\Delta\mu_f$, but one of equations (4.8.20) is redundant, since it applies to a ratio R_e whose value is equal to one minus the number of the other ratios R_e. Any one of equations (4.8.20) may be selected for deletion.

A procedure very similar to the one described in this section was proposed by Clasen (1965). Clasen points out that the method using element potentials is really a higher order improvement of the RAND method in that, at each iteration stage, the set of Lagrangian multipliers χ_e obtained from equations (3.3.34) or (3.3.37), (and these are related to the element potentials μ_e or μ_f) can be improved by the additional solution of equations (4.8.20).

5

GENERAL CONSIDERATIONS AND CONCLUSIONS

5.1 Starting estimates

5.1.1 Estimation procedures

In all the methods of equilibrium computation described in chapters 3 and 4, initial estimates have to be made of some or all of the mole numbers (or mole fractions, or partial pressures) of the species likely to be present in the equilibrium mixture. However, first of all it has to be established which species are likely to be present in the system. Sometimes this knowledge is available from experience or intuition; occasionally an idea of the types of species likely to be present can be obtained from previous calculations. However, very often the information for the selection of species is largely lacking. The most systematic approach would be to consider as possible species all combinations of the elements that are specified for the system in the form of molecular compounds that are known to exist, or, more realistically, in the form of compounds for which thermodynamic data are available in the pressure and temperature regions considered.

As a first step, a decision must be made as to the accuracy with which the equilibrium composition is to be specified. For example, in systems in which N_2 and O_2 are present, NO is a possible combination of elements into a known species. However, the value of the equilibrium constant for the formation of NO from N_2 and O_2 at low temperatures is so small that even under ideal conditions (i.e. under conditions of stoichiometry and in the absence of any other gases in the system) the maximum mole fraction of NO is usually so much below the accuracy required that for most purposes it can be neglected. Only at very high temperatures, e.g. above about 1500 °K and for very accurate work at temperatures somewhat below 1500 °K, should the equilibrium concentration of nitric oxide be calculated. Other possible products can be analysed in a similar manner. By such considerations, many possible constituent species may be eliminated prior to the solving of the complex equilibrium problems (Kobe and Leland, 1954). Baibuz (1962) gives an analytical criterion for the limitation temperature, that is the upper temperature limit above which a compound is not to be considered in equilibrium computations. This analytical criterion is expressed as a function of the accuracy specified for the particular system at a given pressure.

Having selected the possible species that can be present in significant amounts in the equilibrium composition to be computed, it is necessary to establish some initial estimates of these selected species. In computations of equilibria for large numbers of different systems which are similar in character, it is sometimes possible to obtain initial estimates by extrapolation of equilibrium concentrations from other similar but different systems (Horn and Troltenier, 1963). For systems about which no information is available at all, an initial estimate of the composition vector could be the one wherein all mole fractions are assumed to be equal, that is

$$x_i = \frac{1}{N}. \qquad (5.1.1)$$

It was shown in chapter 1 that an analytical criterion exists for the number of components C in a chemical system. Usually, although not necessarily, the number of components is equal to the number of elements $(C = M)$. From experience or from intuition, it is sometimes possible to predict which species will appear in the largest mole fractions in the equilibrium system. Then it is possible to select the main components of the system (optimized basis of components, see §4.5.3) and a possible selection of initial estimates would be

$$\left. \begin{array}{l} x_c = \dfrac{10}{N+11M}, \\[2mm] x_j = \dfrac{1}{N+11M}. \end{array} \right\} \qquad (5.1.2)$$

In these equations it is arbitrarily assumed that all component mole fractions are equal to each other and that all the derived species mole fractions are equal to each other. For systems in which the elemental abundances differ considerably from each other, it is more fruitful to select mole fractions in accordance with mole numbers of components that are directly determined from the elemental abundances. For instance, in systems formed from the combustion of hydrocarbons and wherein almost complete combustion has taken place, CO_2 and H_2O are predominant species and an initial estimate might be that the number of moles of CO_2 in the equilibrium system is equal to the number of gram atoms carbon in that system, and also the number of moles of water in the equilibrium system is equal to half the number of gram atoms hydrogen in the system.

However, in many instances the above simple rules are not applicable, i.e. they do not lead to convergent iterations, particularly in those methods which utilize Newton–Raphson linearization techniques (see §5.2). Of course, a trial and error procedure can be used to find sufficiently close starting values; however, this is very wasteful of computer time and almost prohibitive if more than two or three variables are involved.

Mingle (1963) has proposed a starting procedure which is claimed to be successful in the solution of many problems. The first step in his procedure is to order the mass action equations by their 'importance' (cf. §2.3); this order is determined by the relative magnitude of the equilibrium constants and of the reactants involved. Then, the unknowns are selected as products which re-appear as reactants in additional equations, so that the system can be approximately solved in succession by equating to zero all uncalculated quantities. This produces a set of non-linear algebraic equations, but each one involving only one additional unknown. This system is

$$F_1 = -(K_x)_1 + G_1(x_{1,0} \ldots, o), \qquad (5.1.3)$$

$$F_k = -(K_x)_k + G_k(x_{1,s}, \ldots, x_{k-1,s}, x_{k,0}, \ldots, o),$$

where the subscripts s refer to the starter values calculated from previous equations and are constants. This set of starter values can then be used, for instance as the input for the Newton–Raphson linearization method. In order to solve the set of starter equations by Newton's method applied to each of the variables as they arise, it is still necessary to assume initial guesses for the variables, although with only one unknown in any particular place in the set the solution is much simplified. Mingle's suggested starting procedure will not guarantee convergence, but since the procedure itself will converge if it is repeated in an iterated process, it may be necessary to apply it two or more times before sufficiently accurate values are found so that Newton's method will subsequently converge. Because Newton's method is much more efficient than the starting procedure, Newton's method should be employed as soon as convergence is assured. A suitable divergence test can be made (see §5.2).

In some methods the starting estimates *must* obey the mass balance equations. In this case, e.g. in Brinkley's original method (§4.2.1), it is necessary to select components and to estimate values of the mole numbers of those components by a direct calculation. In this procedure, the mole numbers of all other species are put equal to zero, at least initially. Thus, the estimation procedure only consists of a selection and is not an estimation. The procedure requires an initial matrix inversion to arrive at the selected estimates.

In the method of successive reaction adjustments (§4.5) the initial estimates must also obey the elemental abundances for the system. Smith and Missen (1968) have used a linear programming procedure for the evaluation of the preliminary values. Equation (2.2.2) is written without the logarithmic terms, thus approximately:

$$G = \sum_i n_i \mu_i^\circ, \qquad (5.1.4)$$

where the n_i must obey the mass balance equations (2.2.4):

$$\sum_i a_{ie} n_i = B_e. \qquad (5.1.5)$$

Then, a standard problem in Linear Programming is formed by the requirement that G be minimized subject to (5.1.5) and to the requirement that all $n_i \geqslant 0$. A solution to this Linear Programming problem may or may not exist; this will be largely determined by the selection of the species to be incorporated in (5.1.4).

A feasible solution is most likely to be obtained when an optimized basis of components is selected for inclusion in (5.1.4).

Clasen (1965) also uses a Linear Programming method to find an initial feasible solution to the chemical equilibrium problem. Clasen puts:

$$n_i = m_i + m_{N+1} \qquad (5.1.6)$$

and rewrites the problem as that of finding values of m_1, m_2, ..., m_{N+1} which satisfy

$$\sum_i a_{ie}(m_i + m_{N+1}) = B_e \qquad (5.1.7)$$

and which minimize the function

$$L = \sum_{K=1}^{N+1} C_k m_k, \qquad (5.1.8)$$

where the C_K $(K = 1, ..., N)$ can be put equal to the μ_i°; $C_K (K = N+1) = 0$. In order to insure that all n_i be positive, m_{N+1} can be made positive by maximizing m_{N+1} in another Linear Programming procedure, wherein C_K $(K = 1, ..., N) = 0$ and $C_K (K = N+1) = -1$; then the function $L = -m_{N+1}$.

In the very complex equilibria studied by the RAND Corp. (Shapiro, 1964b), a generalized technique was developed for eliminating species from the calculations. The systems considered in the RAND investigations were multi-phase systems such as the ones described by Dantzig and De Haven (1962). It is beyond the scope of this book to give a detailed treatment of the RAND methods of simplification of complex systems; suffice it to say that the original problem is replaced by a new problem wherein species are eliminated from phases and replaced by new species with constraint factors and energy parameters. This simplifies the problems to the extent that each species needs to be evaluated in only one phase or compartment of the system and its concentration in all other phases follows after the main equilibrium problem has been solved.

5.1.2 Inclusion of condensed species

One of the direct consequences of the phase rule, as was shown in chapter 1, is that the number of condensed phases is restricted by the number of elements present in the system. In the presence of a gas-phase, the maximum number of condensed species or phases that can be present in the system is equal to $M-1$ (see equation 1.3.20). This restricts the number of condensed species that may occur, but does not restrict the total number of condensed species that can be considered for the solution (cf. §2.4).

In the methods described in chapter 3 any number of condensed phases can be included at the start, and their amounts have to be estimated. In the methods described in chapter 4, it is usually not necessary to estimate how much of a condensed phase is present in the system. The simplest procedure normally is to assume the maximum number of possible solids to be present, and to select those that are most likely to be present. If this information is not available, the solids can be selected by suitable permutation techniques in a trial and error method, and the $M-S$ mass balance equations (see equation 2.4.8) can then be solved for the gaseous equilibrium by means of any of the methods described in chapter 4.

It is then part of almost every computational method to decide on the basis of the equilibria computed with the assumption of maximum number of solids or liquids present, whether or not the selected condensed species are indeed likely to be present. If not, other condensed species should be taken into consideration, or condensed species should be deleted entirely. The basis on which a selection is made, at the end of each iterative cycle in the computation, is the chemical potential of the condensed species in question. This is equivalent to discovering whether or not the vapour pressure of the particular species in question is equal to or smaller than the saturation vapour pressure of this particular condensed species. In many of the methods described in chapter 4, negative amounts of solid can be computed. If this is the case, two courses of action are open, (1) to disregard this particular solid in the next iterative cycle, or (2) to add an imaginary amount of the solid species to the system, thus increasing the elemental abundances and changing their ratios, and to calculate the equilibrium accordingly, discovering in this manner which solids or liquids have to be taken into account in the computations.

Thus, for estimation purposes, the phase rule needs not to be taken into consideration and any number of solids may initially be assumed to be present in the system; the iterative procedure can be arranged so that it will reject one after the other of these condensed species, until their number has been reduced to the maximum allowable or less.

5.2   The convergence of numerical methods of solution

5.2.1   Introduction

It has been shown, in chapter 2, that the two main approaches to the chemical equilibrium problem, although apparently quite disparate, are essentially equivalent. This was done by simply deriving the mass action equation form of the problem from the minimization of Gibbs free energy approach.

The equivalence of the two main approaches is not unexpected. However, close scrutiny of the numerical techniques which are actually employed in the methods of solution described, reveals in its turn similar underlying similarities. In fact all these numerical techniques can be grouped, from the point of view of their convergence properties, into three classes. These will be referred to below as (i) Pure iteration solutions, (ii) Newton–Raphson method solutions and (iii) Differential equations solutions. It is of some interest to look more closely at the convergence properties of these three classes of solution, particularly with reference to their limitations. This section will be devoted to a brief discussion of these points.

5.2.2   Pure iteration solution

This class of approaches includes several of the methods discussed in chapter 4, and is employed for instance, by Brinkley (1947). Those methods making use of convergence forcing, such as the Nested Iterations method (Storey and van Zeggeren, 1967), also fall into this class. Thus, this technique is found employed mainly in the non-linear equation form of the problem.

In all these cases, the problem is presented as a set of N non-linear equations in N variables. These are of the form

$$x_i = \Phi_i(x_1, \ldots, x_N), \qquad (5.2.1)$$

where $i = 1, \ldots, N$.

The solution method adopted is that of substituting approximate values y_i^r into the right-hand side of the equations (5.2.1) which are then used to calculate the next, and, it is hoped, a better approximation y_i^{r+1}. This may be written as

$$y_i^{r+1} = \Phi_i(y_1^r, ..., y_N^r). \tag{5.2.2}$$

The superscript r refers here to the iteration. For the desired root x_i:

$$x_i = \Phi_i\{x_1, ..., x_N\}. \tag{5.2.3}$$

In order to discuss the convergence properties of this solution method we proceed as follows:

Let the y_i^r be a given estimate of the desired root x_i such that

$$y_i^r = x_i + \delta_i^r, \tag{5.2.4}$$

where δ_i^r is small. (A more precise meaning will be given to the term 'small' below.) Substituting for y_i^r in equation (5.2.2) leads to

$$y_i^{r+1} = \Phi_i\{x_1 + \delta_1^r, ..., x_N + \delta_N^r\}. \tag{5.2.5}$$

Expanding the right-hand side of (5.2.5) as a Taylor series in the δ_i^r about the x_i leads to

$$y_i^{r+1} = \Phi_i\{x_1, ..., x_N\} + \sum_l \left(\frac{\partial \Phi_i}{\partial y_l}\right) \delta_l^r \tag{5.2.6}$$

to first order in δ_i^r, where l is a dummy index for $i(l = 1, ..., N)$. Here it has been assumed that all the Φ_i are well behaved, so that the expansion converges, and that the δ_i^r are small enough to make the expansion to first order a good approximation.

Putting now

$$y_i^{r+1} = x_i + \delta_i^{r+1} \tag{5.2.7}$$

so that δ_i^{r+1} is the error in the next approximation to the root, then, using equations (5.2.3) and (5.2.6) leads to:

$$\delta_i^{r+1} = \sum_l \left(\frac{\partial \Phi_i}{\partial y_l}\right) \delta_l^r \tag{5.2.8}$$

or

$$\delta_i^{r+1} = \sum_l D_{il} \delta_l^r. \tag{5.2.9}$$

Define the quantitites ρ_{lu} and π_{iu} as:

$$\sum_l D_{il} \rho_{lu} = \lambda_u \rho_{iu}, \tag{5.2.10}$$

$$\sum_i \pi_{iu} D_{il} = \lambda_u \pi_{lu}, \tag{5.2.11}$$

i.e. as the eigenvectors of the matrix D_{il}, the λ_u being the eigen-values. The index u is a dummy index for i ($u = 1, ..., N$). Let the ρ_{lu} and π_{iu} be normalized, i.e. let

$$\sum_i \pi_{iu} \rho_{lv} = \delta_{uv}, \qquad (5.2.12)$$

where v is a dummy index for u ($u,v = 1, ..., N$) and where δ_{uv} is the Kronecker delta.

For our purposes we can (Higman, 1955) now expand

$$D_{il} = \sum_u \lambda_u \rho_{lu} \pi_{iu} \qquad (5.2.13)$$

and

$$\delta_l^r = \sum_u C_u^r \rho_{lu}. \qquad (5.2.14)$$

The expansion for δ_l^{r+1} can now be found by substituting equations (5.2.13) and (5.2.14) into equation (5.2.9) and using the normaliza-tion to yield

$$\delta_i^{r+1} = \sum_u C_u^r \lambda_u \rho_{lu} \qquad (5.2.15)$$

and after r^1 further iterations

$$\delta_i^{r+r^1} = \sum_u C_u^r \lambda_u^{r^1} \rho_{lu}, \qquad (5.2.16)$$

where the λ_u are raised to the power r^1. Thus the iteration method will only converge (i.e. the successive iterates y_i^r will only tend to x_i as the iteration progresses) if *each* λ_u satisfies the condition

$$|\lambda_u| < 1. \qquad (5.2.17)$$

This condition is unlikely to be satisfied by chance, and be-comes even less likely as N increases. This explains why such extreme care must be taken in those methods using this approach to arrange the equations properly on the correct set of variables. What the rearrangement accomplishes, in fact, is the formulation of a set of equations for which conditions (5.2.17) will be satisfied. In practical cases the arrangement of the equations appears to be of much greater importance than an accurate initial estimate.

Convergence forcing methods, in which new estimates are found by means of expressions such as:

$$(y_i)_{\text{new}} = \lambda y_i^r + (1 - \lambda) y_i^{r-1} \qquad (5.2.18)$$

instead of using the iterates y_i^r directly, involve one or more, normally hand picked, parameters λ (see e.g. Hilsenrath *et al.*

1959). Iterations of this type are again effectively of the form of equation (5.2.1) and subject to the same analysis as above. By suitably adjusting the value of the parameter, however, it is possible to control artificially the value of one of the $|\lambda_u|$ for each parameter used, and thus to some extent force convergence even in many-variable iterations. The only way, however, to guarantee first-order convergence in an N-variable iteration is to employ N parameters, which is done in the Newton–Raphson method.

5.2.3   Newton–Raphson method

The Newton–Raphson (or Newton's) method is probably the most popular (and is certainly the best known) method of finding numerically the roots of a set of non-linear equations. It is used in a number of the methods treating the chemical equilibrium problem described in earlier chapters of this book, viz. in the RAND method (§3.3.2), in Brinkley's method (§4.2.2) and in the NASA method (§4.3.1). For a comparison of these three methods and for a discussion of their computational equivalence see Zeleznik and Gordon (1960, 1968).

A detailed treatment of Newton's method is beyond the scope of this book. However, the treatment can be taken far enough, without undue complication, to provide some insight into the properties of Newton's method as observed in practice.

The equations are here presented in the form

$$\Phi_i\{x_i, ..., x_N\} = 0, \tag{5.2.19}$$

where the desired root, denoted by $\{x_i\}$ actually satisfies these, and an approximation to the desired root, $\{y_i^r\}$ leaves a small 'residual' F_i on the right-hand side when substituted into the Φ_i, i.e.

$$\Phi_i\{y_1^r, ..., y_N^r\} = F_i. \tag{5.2.20}$$

To derive the algorithm employed by Newton's method, we proceed as follows:

$$y_i^r = x_i + \delta_i^N \tag{5.2.21}$$

so that equation (5.2.20) becomes

$$\Phi_i\{x_1 + \delta_1^r, ..., x_N + \delta_N^r\} = F_i. \tag{5.2.22}$$

The left-hand side of equation (5.2.22) can now be expanded as a Taylor series about x_i (where we assume, as usual, that the Φ_i are such as to make the expansion permissible). To first order in the δ_i^r we have then

$$\Phi_i\{x_1, \ldots, x_N\} + \sum_l \left(\frac{\partial \Phi_i}{\partial x_l}\right)\delta_l^r = F_i. \qquad (5.2.23)$$

Now write D_{il} for the elements of the matrix of partial derivatives

$$[D_{il}] = \frac{\partial \Phi_i}{\partial x_l} \qquad (5.2.24)$$

and let D_{li}^{-1} represent the elements of the inverse of the matrix D_{il}. In this case equation (5.2.23) reduces to

$$\sum_l D_{il}\delta_l^r = F_i \qquad (5.2.25)$$

so

$$\delta_l^r = \sum_i D_{li}^{-1}F_i. \qquad (5.2.26)$$

Hence a better estimate of the root $\{x_i\}$, which will be denoted by $\{y_i^{r+1}\}$, is given by

$$y_i^{r+1} = y_i^r - \sum_l D_{li}^{-1}F_l. \qquad (5.2.27)$$

This is the basic algorithm to Newton's method (cf. §4.2.2).

In order to determine the behaviour of the errors in the successive approximations to the root, first rewrite equation (5.2.27) as

$$y_i^{r+1} = y_i^r - \sum_l D_{il}^{-1}\Phi_l\{x_1^r, \ldots, x_N^r\}. \qquad (5.2.28)$$

Putting now

$$y_i^{r+1} = x_i + \delta_i^{r+1} \qquad (5.2.29)$$

and expanding the Φ_l on the right-hand side of equation (5.2.28) to *second* order in the δ_i^r, leads to

$$\delta_i^{r+1} = \delta_i^r - \sum_l D_{il}^{-1}\left\{\sum_i D_{li}\delta_i^r + \frac{1}{2}\sum_{u,v}\left(\frac{\partial^2 \Phi_l}{\partial x_u \partial x_v}\right)\delta_u^r \delta_v^r\right\} \qquad (5.2.30)$$

or

$$\delta_i^{r+1} = -\frac{1}{2}\sum_{l,u,v}\left\{D_{il}^{-1}\left(\frac{\partial^2 \Phi_l}{\partial x_u \partial x_v}\right)\right\}\delta_u^r \delta_v^r. \qquad (5.2.31)$$

Thus, the amplitude of each successive error is, roughly speaking, proportional to the *square* of the amplitude of the preceding error, as opposed to being linear in the amplitude of the preceding error, as in the pure iteration method. Newton's method is thus a second-order method. If, in any given application, Newton's method converges, and this is by no means guaranteed, then its rate of convergence will be in most cases substantially faster than that of the pure iteration method.

The main difficulty found in practice with the ordinary Newton's method, apart from the need to calculate, and invert, the matrix of first derivatives D_{il} at each step, lies in the requirement that the starting estimate be rather a good one if the method is to converge. The above analysis shows that, if convergence is achieved, it is rapid. On the other hand, a poor starting approximation largely invalidates the analysis since third and higher order terms can no longer be neglected.

Almost identical considerations apply to those minimization methods for solving the chemical equilibrium problem which attempt to reach the minimum in a single step (e.g. the RAND method). Such methods, on being subjected to analysis in the above manner, normally lead to equations of the form of (5.2.30) above (cf. Brandmeier and Harnett, 1960).

5.2.4 Differential equations solution

The third main type of numerical method used consists basically of the conversion of the problem into the solution of a set of simultaneous first-order differential equations. These methods make no attempt to reach a particular point of interest in a single bound, but rather deliberately eschew the direct approach, and reach the solution required in a series of small steps.

Thus, this class of approaches to the solution of the problem includes the first-order steepest descent method (§3.3.1), Naphtali's gradient method (§3.4.1) and Storey's survey method (§3.4.2). In all of these, the operating equations are effectively a set of first-order differential equations specifying rates of change of the n_i values with respect to some parameter.

The main disadvantage of methods of this type is the length of time normally taken to reach a solution. The crude integration methods which have so far been used force the use of a small step size to minimize accumulated errors in the solution. In addition, it is normally necessary to invert a matrix at each step. The first of these difficulties can be remedied somewhat by the use of higher-order integration methods such as the well known fourth-order Runge–Kutta method, or possibly one of the predictor-corrector methods, although there does not appear to be any evidence available on these points in the literature.

The chief advantage of the differential equations approach is that it will always converge (eventually) to the solution, and do this, in the case of the first-order steepest descent methods, no matter how bad the starting estimates may be. Indeed, as a result of this, the differential equations approach may in the long run lead to the solution faster than either the pure iteration or Newton approach when a completely unknown system is being considered for the first time.

It is this property of ultimate convergence in the differential equations approach that has to be sacrificed to gain the additional speed of the other two approaches. In increasing the speed with which the solution may be reached, one loses the guarantee that an answer may be obtained at all.

Two methods described in chapter 4, viz. Scully's trial and error method (§4.4.1) and Villars' original method of successive reaction adjustments (§4.5.1), also belong in this general class of methods in which the equilibrium is approached in a series of steps. Although these two methods are not described in terms of differential equations, they are in effect based on finite difference equations corresponding with first-order differential equations.

5.3   Consideration of condensed phases and of non-ideality

It is probably fair to state that nearly all the methods for the computation of chemical equilibria, described in chapters 3 and 4, were originally developed by their inventors for application to ideal gaseous mixtures. However, almost all problems encountered in practice for which a more elaborate than manual calculation is required, are problems encountered in systems wherein condensation of some species does or can take place under certain conditions, specified for the problem, and/or in systems at pressures which would make treatment as ideal systems introduce unallowable errors.

In this section the various methods will be compared as to the ease with which condensed phases and non-ideality can be taken into account in the methods.

5.3.1 Inclusion of condensed species

For purposes of the discussion in this section it will be sufficient to consider only pure condensed species, and in particular pure solid species. The treatment for inclusion of mixed condensed phases is always completely analogous to that for inclusion of pure condensed species.

As was demonstrated in §2.4, application to systems containing pure condensed species is substantially easier for methods based on minimization of G or A (see chapter 3), than for methods based on the solution of sets of non-linear equations (see chapter 4). The main reason for this is that the basic mathematical framework of minimization methods is independent of the form of the chemical potential used, whereas in the non-linear equation methods the set of equations required for the solution depends on the number and types of condensed species assumed to be present. It was shown in §2.4 that the mass balance equations required for the computation of the gaseous equilibrium decrease in number and the mass action equations increase in number, both by the number of additional condensed species considered. Furthermore, the selection of mass action equations and mass balance equations appearing and disappearing is dependent on the selection of the particular types of condensed species assumed to be present in the equilibrium to be calculated.

Another major difference between the two groups of methods is that in the minimization methods the amounts of condensed species have to be estimated, whereas in the non-linear equation methods only the types of condensed species need to be selected, at least in principle (cf. §5.1.2).

In addition to the above-mentioned general differences, there are a number of more subtle differences in the methods, at least as far as inclusion of condensed species is concerned. These differences are described in somewhat greater detail below.

(1) *Optimization Methods:* For none of the general optimization

techniques, described in §3.2, has the ideal gas treatment been extended by the various authors to include condensed phases. Such extensions would be interesting, although it is very likely that they would require extremely careful programming. It is easy to envisage that, during the course of the optimization procedure, several of the condensed species might vanish simultaneously and the searches would become divergent or at least very inefficient. Since all three general methods in §3.2 are, already, quite inefficient, their extension to include condensed species would probably not be a very fruitful exercise.

The steepest descent methods, described in §3.3, are much better suited for consideration of condensed species. In particular the first-order search, although very slow, is capable of handling large numbers of condensed species. Even though most of these species might tend to disappear as equilibrium is approached, the fact that negative mole numbers are precluded indicates that no problems will be encountered in calculating mole numbers that persist in going negative. The second-order RAND procedure (see §3.3.2 for details) works very satisfactorily for systems where one can predict *a priori* which condensed species will be present in the chemical equilibrium to be computed. It is usually best to include as many condensed species as possible at the start of the computation, and then to remove, in some systematic sequence, those species that, on iteration, persist in going negative. This is also the procedure followed by White *et al.* (1958) for gaseous species that persist in going negative. Inclusion of too many condensed species may, on occasion, prove troublesome in that G would become substantially linear. If only condensed species were present, of course, so that G would be completely linear, G would not have a minimum, in the sense assumed, at all. Oliver *et al.* (1962) provide a single criterion to decide whether a particular pure species, not included at the start, should be included at any given stage in the computer program. Their criterion is based on the equation, derived for gases in chapter 2 (see equation (2.3.9)), but is also applicable to those condensed species (j) that are assumed to be present in the system, viz:

$$\mu_j^0 - \sum_c \mu_c a_{cj} = 0. \qquad (5.3.1)$$

For those condensed species that were not included at the start of the computation, the left-hand side can be evaluated after each iterative cycle; when negative, the condensed species j has to be included prior to the next iteration. It should be pointed out here that the criterion

$$\mu_j^0 - \sum_c \mu_c a_{cj} \geqslant 0 \qquad (5.3.2)$$

is equivalent to the condition that, at equilibrium, the partial vapour pressure of a pure condensed species is always less than or equal to the saturation vapour pressure of that species (cf. §5.1).

As was mentioned in §3.4.1, Naphtali's gradient method encounters difficulties when large numbers of possible condensed species are included at the outset of the computation, since there is no implicit procedure for preventing the n_s from becoming negative. Of course, as Naphtali points out, scale parameters can be introduced, but such artificial means tend to slow down the computation *without guarantee of convergence*. The only alternative procedure is to select permuted combinations of all possible condensed species until a combination is found which does not lead to any of the n_s becoming negative (as outlined in §5.3.2). However, this is often so slow as compared to other methods, that it is to be done only in instances where computer memory is large and cheap, and computational speed is extremely rapid.

The survey technique (§3.4.2) can handle condensed species as indicated earlier in this section in the discussion of the RAND method. Starting from an equilibrium point and going to the next, the possibility of absence or presence of each conceivable condensed species must be checked on, with equation (5.1.2), after the new equilibrium has been computed. The fact that, in surveys, the path followed by the computations is an equilibrium path, ensures an approach whose physical significance makes the appearance and disappearance of condensed species a physically predictable phenomenon; this facilitates the checking procedure considerably.

(2) *Mass Action Equation Methods:* As was mentioned above, many of the non-linear equation methods require reduction (by S)

of the mass balance equations necessary for the actual equilibrium computation, and increase (by S) of the mass action equations. This is the case for the Brinkley method (§4.2), for the trial and error methods (§4.4), for the NOTS method (method of successive reaction adjustments, see §4.5), for the nested iterations method (§4.6), and for all the miscellaneous methods (§4.7), although it is only explicitly shown for the nested iterations method. In all cases the treatment prescribed by Boll (1961) for the Brinkley method can be applied equally well, after suitable modification, to the other methods. Thus: prior to each iteration, estimate whether a condensed species will be present and, after each iteration, check whether this was a valid estimate. In these methods, S may only be taken to be equal to or less than $M-1$; in other words: the number of condensed species should not exceed that allowed by the phase rule. In fact, in the Brinkley method all condensed species have to be selected as components; they may not be considered as derived species for the purpose of computation. In the NOTS method, condensed species are considered to be primary derived species.

The only two exceptions to the rules and restrictions of the above paragraph are afforded by the NASA and element potential methods. In the NASA method, condensed species need not be components; in fact, only gaseous elements are allowed to be components. Also, as many condensed species as desired can be included in the computational procedure; those incorrectly assumed to be present will *gradually* become zero. The only disadvantage may be considered to be that the total number of implicit equations to be solved is increased by S, but this is a small price to pay for the convenience with which condensed species can be incorporated in the method. The other exception to the rule of the above paragraph, i.e. the elemental potential method, is one in which corrections not to the mole numbers, but to element potentials are computed, and these corrections can logically be used, in equation (5.3.2), to predict whether or not a given condensed species will be present. As was pointed out in §4.1, the discussion of element potentials might just as conveniently have been placed in chapter 3.

5.3.2 Treatment of non-ideality

In §1.4, a summary is given of the equations of state describing the interdependence of pressure, volume and temperature of non-ideal systems. The discussion here will be limited to non-ideal gaseous systems, although most of the comments made are equally pertinent for non-ideal condensed phases.

In all the methods described in chapter 3, the minimization of the Gibbs free energy G of a system really involves the minimization of a sum of functions of the mole fractions of the species in that system, as expressed in terms of the partial molar Gibbs free energies, or chemical potentials, μ_i. For non-ideal systems these μ_i can be expressed as (equation 1.4.15):

$$\mu_i = \mu_i^\circ + RT \ln f_i. \qquad (5.3.3)$$

For systems wherein the f_i can be expressed in terms of non-ideal and ideal parts, a separation can be effected, e.g. (cf. 1.4.18):

$$\mu_i = (\mu_i)_{\text{ideal}} + RT \ln \gamma_i \qquad (5.3.4)$$

so that then the Gibbs free energy G can be expressed as:

$$G = G_{\text{ideal}} + G^E, \qquad (5.3.5)$$

where G^E is a thermodynamic excess function, for which an analytical expression is then available:

$$G^E = \sum_i n_i RT \ln \gamma_i. \qquad (5.3.6)$$

The γ_i are, in general, functions of all $n_i (i = 1, ..., N)$, and for some equations of state applicable to some gas mixtures under some conditions, an analytical function for γ_i can be obtained. Then, differentiation with respect to the n_i becomes possible and the minimization procedures of chapter 3 can treat the non-ideality condition (5.3.6) with relative ease (see e.g. Boynton, 1963). In some instances, it is not possible to express G^E in terms of γ_i, as in equation (5.3.6), but as long as any differentiable expression can be found, a minimization solution can be obtained. The only published information on such an extension of a minimization method to non-ideal systems is an article by McGee and Heller (1962) whose used the second-order (RAND) method for their treatment.

In all the methods described in chapter 4, the same procedure for inclusion of non-ideality can be followed: initial estimates are made, and the first step in the iteration procedure is carried out, assuming ideal behaviour. Subsequently, a correction is made for non-ideality via the calculation of a non-ideality factor b_j in the mass action equations, see equation (4.2.35), and step 2 in the iteration procedure is carried out, this time incorporating the non-ideality factor computed after step 1. This procedure, while simple and applicable to all methods, is sometimes not successful, particularly at extremely high densities where corrections become very large. For instance, in the computation of explosive properties, factors b_j of 1000 or larger are not uncommon.

A better procedure for considering non-ideality is to incorporate it in each step in a typical Newton–Raphson procedure, such as is applied to the Brinkley method. This implicit procedure is described in §4.2.4. It is very frequently successful and still quite simple. More tedious procedures have been developed for the implicit consideration of non-ideality; e.g. Michels and Schneiderman (1963) presented a treatment of the NASA method with application of a complete virial equation of state (equations 1.4.4 or 1.4.5).

It is very likely that better methods can be developed for the computation of chemical equilibria in non-ideal systems For instance, the trial and error methods, the nested iteration method, and the NOTS method, would all be amenable to treatments different from and probably better than those described above. It is to be hoped that further investigations of these problems will be carried out.

5.4 The effect of errors in the thermodynamic data

5.4.1 Introduction

In view of the extent to which experimentally derived data is employed in the computation of chemical equilibria, and of the often substantial errors inevitably present in such data, it is remarkable how little work has appeared on this subject.

It is, of course, always possible to attack the problem directly. Having completed a given computation, a small change can be made in a suspected parameter, and the computation repeated to see what the effect on the equilibrium composition will be. Unfortunately, for large scale problems, or problems for which a number of parameters are under suspicion, this approach becomes far too expensive in computer time and a more systematic approach becomes necessary.

One approach is to attack the problem via the mass action equations and use a straightforward first-order expansion in the changes in the parameters of interest (and, of course, in the corresponding changes in the equilibrium n_i values) (Neumann, 1962). This would seem to be a perfectly satisfactory approach for cases when temperature and pressure are held constant, and, provided no unusual difficulties arise, such as a singular matrix of partial derivatives, will allow all the partial derivatives of interest to be found in a single computation. The assumption of small errors only is not unreasonable, since a large error in a parameter is presumably not very likely to escape detection.

This approach, however, becomes less satisfactory if changes in temperature and pressure are to be considered. Both of these quantities are assumed constant in the derivation of the mass action equations from the minimization of G, for example, and the problem at once becomes exceedingly complex.

5.4.2 The second minimization approach

The most frequently encountered form of the chemical equilibrium problem is that of ideal phases at constant temperature and pressure. Even in this simplest case, there are three possible sources of error in the data to be employed in the computation. First, the μ_i° values employed may be subject to experimental error. Second, the elemental abundances may be subject to purely numerical errors. Finally, the coefficients of the mass balance conditions may be subject to errors, again of a purely numerical form.

Since errors in the coefficients of the mass balance equations are presumably extremely unlikely (although they also may be treated by the technique discussed in this section) they need not be considered here. Also, errors in the elemental abundances are unlikely, since, as for the coefficients of the mass balance equations, the mass balances are normally determined by the setter of the problem and are thus, as a rule, subject to arithmetic errors only. However, situations can arise when the elemental abundances of an equilibrium composition will have been determined experimentally, and may be under suspicion. Ideally, in this case, it is desirable to be able to calculate, given say the probable error in the suspected elemental abundance (s) the corresponding quantities for the computed equilibrium composition. In the case when several pure condensed species may be present, errors in the elemental abundances can affect not only the quantity of the individual species present, but can even change the species which appear.

So far, no material appears to be available in the literature aimed specifically at the calculation, say, of the standard deviation of the equilibrium composition produced by uncertainties in the elemental abundances. The problems involved are indeed very complex. The most hopeful approach to this problem, and to the more urgent problem of errors in the partial molar free energies, would seem to be through the second minimization approach employed in §§2.6 and 3.4.2, which, in spite of its limitations, allows errors of various types, including errors in temperature and pressure, to be discussed.

The discussion of §3.4.2, which will not be repeated here, provides a means of obtaining an accurate estimate of the effect on the equilibrium composition of a small change or error in the elemental abundances of an ideal system at constant temperature and pressure, without repeating the entire equilibrium computation. It can be generalized without too much difficulty to the consideration of systems with variable temperature and pressure (van Zeggeren and Storey, 1969). In spite of this, its convenience, in that only a relatively small matrix need be inverted, is due entirely to the fact that only ideal systems are considered. Although the second minimization approach can be used on non-ideal systems, in this case, since the second partial derivatives of G with respect to the n_i do not have the same simplicity of form as for ideal systems, the set of simultaneous equations to be solved is much larger. The total number of simultaneous equations involved is of the order of the number of species rather than, as in the simpler case, of the order of the number of elements. Apart from this, complications arise in systems including condensed species in those cases where the error causes a different set of solids to be present at equilibrium to that which would be present if the error had not been made. In view of these points, it is not altogether surprising that the general problem of obtaining the probable errors in the equilibrium composition from probable errors in the elemental abundances should not yet have been considered.

The same difficulties are inherent in the discussion of the most important of the three sources of error, that of errors in the partial molar Gibbs free energies (μ_i°). A means of treating this problem, again for ideal gas systems at constant temperature and pressure, has already been given in §2.6, and the reader is referred to that section for details of the application of the second minimization technique in this case. It should be noted that the treatment of this type of error is not quite the same as that of errors in the mass balance conditions. The errors in the μ_i° affect the minimized function itself, whereas changes in the elemental abundances affect the side conditions involved. Here again, it would be desirable to be able to calculate, say, probable errors in the solution from the probable errors in the partial molar free energies, but the problem does not appear to have been considered, although some useful general rules have been formulated (Shapiro, 1964a). Criteria for halting computations are sometimes based directly on estimated accuracy of the μ_i° (Naphtali, 1960). Clearly, a great deal of further work in this area is required before this aspect of the chemical equilibrium problem reaches a satisfactory state.

5.5. The comparison of computational methods

5.5.1 Introduction

Earlier chapters of this book have not only discussed the problems involved in the computation of chemical equilibrium compositions, but have provided a sufficiently detailed description of a number of methods for the numerical solution of the problem to allow the reader to proceed directly to carrying out his own computations.

In fact, by far the greater part of the book has been devoted to the description of the various types of methods currently available in the published literature. It is natural, therefore, to wonder whether it is in fact necessary to discuss so many, and whether it is not possible to point to one particular method as the 'best' and ignore the rest.

One justification for this book is that there is no simple answer to this question. The main difficulty is to decide what criterion is to be used in selecting the 'best' method. In the published literature on the field (e.g. Piehler, 1962; Brandmeier and Harnett, 1960) it is often implied that the best method is that which is computationally the fastest, i.e. that which yields the desired equilibrium from a given starting estimate in the minimum time. In the past, such a criterion has often been dictated by the relatively high cost of computer time, which made the fastest running method the most economic, regardless of convenience to the user. On the other hand, cases can arise where the most convenient method and not the fastest can be considered the best, particularly in such cases where computer time is relatively cheap compared to the user's time. Both of these criteria can be seen to suggest an underlying criterion which is in fact economic. However, if total computation cost (machine + data + personnel) is used to choose the best method, then the particular installation to be used, and its costing procedures, at once become the most important factors in any comparison, and a general overall comparison is simply not possible.

In §5.5.2 a discussion is presented of the many pitfalls encountered in attempting to make a direct comparison of computational speeds of the various methods.

The general considerations in §§5.1, 5.2, and 5.3 provide other criteria for a comparison of the various methods; these will be reviewed in §5.5.3, and general recommendations will be given to the reader as an aid in selecting that particular method that will be best suited to the problem to be solved.

5.5.2 Comparison of computational speeds

A traditional form of making comparisons of various methods is to choose a standard and reasonably realistic problem (such as the combustion of propane with air) and then simply to try the various methods to be compared on this problem, to see which method solves the problem to the required accuracy in the minimum time. However, this superficially simple procedure is beset with many problems. Some of these should be obvious to most users (but not necessarily to their programmers). Conversely, some of the more subtle problems may well be obvious to a relatively experienced programmer, but not, in many cases, to the physical chemist or chemical engineer requesting the comparison. This is not to say that the traditional direct comparison approach is not without value or interest, but simply that its results must be interpreted with a certain amount of caution.

First of all, it is unrealistic (as well as perhaps a little unfair) to compare, say, a method which is intended specifically for non-ideal systems with one which is not, or one intended for large systems with one intended for small. A second difficulty is that, even within methods which are specifically aimed at treating the same class of problem, such as an ideal gas system at constant temperature and pressure, the particular test problem chosen can have a substantial effect on the relative speed of two methods. While the second difficulty can be partially overcome by comparing the results of two or three substantially different test problems, the first difficulty cannot be remedied so easily. It is perfectly possible in extreme cases to select a problem which can be solved easily by one method but not at all by another, simply because the other method was never intended for problems of this type. This does not necessarily imply that the other method might not be the faster of the two for problems within its own sphere, as indeed it often will be.

The third of the more obvious problems finds its origin in the influence of the starting estimate on the speed of solution. Here, the main difficulty arises in the comparison of two methods which may have differing restrictions on the starting estimate. For example, the one method may require that the starting estimate satisfy the mass balance requirements of the solution desired, while the other method may require a starting estimate that satisfies the mass action equations. (The only starting estimate which can simultaneously satisfy both is of course the solution.)

Even for two methods with unrestricted starting values, or which employ starting estimates with the same restrictions, the actual starting value can have an important effect. It is not unusual for one of the methods to behave perfectly normally, while the other, with the same starting value, may not converge to a solution at all, although the latter, in such cases when it does converge, may be the faster of the two. Thus, although the use of a variety of starting estimates allows the influence of the starting estimate to be eliminated to some extent among methods with unrestricted starting estimates (or starting estimates with identical restrictions) the problem of comparing methods with radically different restrictions on the starting values does not yet appear to have been considered.

Turning now to the mechanics of the computation itself, several further difficulties can be seen to beset the execution of a direct speed comparison. These are perhaps less obvious than those described above, but can have at least as much influence on any results obtained. It is assumed now that the comparison, as with the vast majority of equilibrium computations carried out today, is to be executed on a digital computer of some kind (only one article has been found to deal with analogue computation: McEwan and Skolnik, 1951).

(a) First, one must take into account the size and type of the computer itself. Some methods, which may work very well on the computers with very large rapid access memories frequently found in Universities and Government laboratories, may be exceedingly awkward, if not impossible, for the often much smaller computers found in industrial laboratories. Apart from this, the *relative* speed with which a computer carries out such operations as addition and multiplication during the course of a computation can also, of course, affect the relative speeds of two different methods. The relative speeds of such operations (provided they be due to the basic hardware) can, however, be compensated for by running comparison tests on a suitable variety of computers. In doing so, it should be borne in mind that two computers by the same manufacturer, which may be ostensibly the same model, may differ in relative internal speeds from one installation to another as a result both of the properties of the individual installations as well as of a variety of technical reasons. Improvements in manufacturing processes can also have an effect if one computer is of substantially earlier vintage than the other.

(b) A second, and in many cases more important, consideration is the effect of the properties of the overall system under which the computer in question is operated. In particular, the language in which the programme is written is of major importance. Both of these, but especially the latter, can have substantial effects on the relative speeds of two methods. The problem becomes acute if two methods are to be compared by means of programmes written in high level languages such as FORTRAN or ALGOL. The particular language used, or, more realistically, the particular subset of the language implemented can, again, favour one method over another, often as the result of the structure of the compiler used. Thus, realistic comparisons must take into account the effect of the system, the language and, where relevant, the compiler.

(c) Finally, one must take into account the influence of the programmer himself, whose style and experience may bias the result of a comparison just as they can have an enormous influence on the speed with which a programme eventually runs. A good programmer, who thoroughly understands the system and (if he or she is using a high level language) the compiler employed, can frequently produce a programme which runs an order of magnitude or more faster than the equivalent programme produced by someone less skilled. While such effects can be studied by having several programmers, at various levels of experience, run comparisons, the number of factors influencing the 'simple' direct comparison of the speeds of the various computational methods is once again increased. Even if such a series of tests is attempted, however, it will still be necessary to take into account the fact that the level of experience of the programmer may well alter during the course of the test.

5.5.3 Summary of properties

The previous sections have raised briefly some of the factors which must be considered in any attempt to find the 'best' of the numerical methods for the computation of chemical equilibria.

Enough detail has been given, however, to make it clear that even in what is ostensibly the simplest case, that of direct comparison of the *speed* of the different methods, an elaborate series of tests involving a considerable variety of factors is necessary if even reasonably objective results are to be obtained. The results of such tests will, in any case, prove difficult to sort out, although a number of statistical techniques are available for considering this type of problem.

The material presented in §§5.1, 5.2 and 5.3 provides several other criteria for the comparison of the various methods. Thus, methods might be compared on the basis of the degree of closeness to the actual solution, required for the starting estimates, in order that that solution may be obtained. Also, methods could be compared as to their likelihood of convergence, for one or a series of given starting estimates. Finally, since all methods differ in the ease with which condensed phases and non-ideality can be taken into account, this ease could also be used as a criterion for comparisons.

Apart from the technical problems involved in a comparison, the additional problems introduced by the variety of criteria for defining the 'best' method, increases the complications of any kind of assessment to a point where the concept of a 'best' method becomes virtually meaningless. This suggests that, in fact, there is *no* single best method, and, even if there was, it would be virtually impossible to agree on which it was. A similar conclusion has been reached, although by a somewhat different route, by Zeleznik and Gordon (1968) in a most interesting recent review.

Since detailed comparison of the methods is thus not feasible, the most sensible way of providing the reader with some form of guidance in the use of the techniques described in earlier chapters is probably to summarize the methods from the point of view of the types of problems (and computer) to which they seem best suited. Such a summary, where all the various criteria are taken into consideration, is presented in table 5.5.1 for all the methods of chapters 3 and 4.

The problem conditions considered in the table are:
   (i) Whether the computer to be used has a limited rapid access memory or not;

(ii) Whether a good or a poor initial estimate is available;

(iii) Whether a single computation is to be carried out, or routine computations on a familiar system are to be executed, or a survey over a range of initial compositions is required;

(iv) Whether non-ideal systems are to be considered;

(v) If ideal systems only are to be considered, whether gaseous species only are present, or whether condensed species are also to be included;

(vi) Whether such parameters as temperature or pressure are to be variables.

Since most of the methods *can* be used (if occasionally somewhat unsatisfactorily) under most conditions, the stars in the table are intended only to suggest that the starred methods are more suitable under the given conditions than those not starred. The recommendations in the table are based as far as possible on the systems that the originators of the methods discuss in their original work as well as on the work of their successors, as described in the relevant sections of chapters 3 and 4. The recommendations are inevitably coloured by the authors' own experiences with a variety of the current methods (although it is fair to say that the authors are perhaps somewhat critical of all the methods published so far, including their own).

Table 5.5.1, then, is intended simply as a rough and rather makeshift guide to assist the reader in selecting suitable methods for his particular problems until such time as he has enough experience to choose (or develop) a best method of his own.

In order to provide the interested reader with a means to compare the various methods, and to test each method and its appropriate computer program for accuracy, Table 5.5.2 is composed of the results of computations for a typical example. The example selected for the purpose is the same as that chosen by a number of other authors (Damköhler and Edse, 1943; Kandiner and Brinkley, 1950; and Scully, 1962), viz. the combustion of propane with air, to give decomposition products, at equilibrium, at 2200 °K and 40 atm. pressure. The only system parameter considered in the table is the molar ratio $\mathcal{R}$ of air

($1\,O_2 + 4\,N_2$) to propane ($1\,C_3H_8$). All data required for the solution of the example problem were taken from the most recent JANAF tables (Stull, 1965).

TABLE 5.5.1 *Comparison of methods of equilibrium computation*

In the table, a star denotes that the method is applicable under the given problem conditions.

	1	2	3	4	5	6	7	8	9
Direct Search (3.2.1)		*	*			*		*	*
SUMT (3.2.2)			*			*			
Linear Programming (3.2.3)		*	*			*			
1st Order (3.3.1)	*	*	*			*	*	*	*
RAND (3.3.2)			*	*	*	*	*	*	
Naphtali (3.4.1)	*	*	*			*		*	
Survey (3.4.2)				*	*	*			*
Brinkley (4.2.1)	*		*			*			*
Brinkley–Neumann (4.2.2)			*	*	*	*	*	*	
NASA (4.3.1)			*	*	*	*	*	*	*
Component Reduction (4.4.1)	*	*	*			*			
NOTS (4.5.4)	*			*		*	*		
Nested Iterations (4.6)	*	*	*	*	*	*			*
Brute Force (4.7.1)	*		*			*			
Element Potentials (4.8)			*			*	*		

Key to problem conditions

1	Restricted computer memory	2	Poor initial approximation
3	Single computation	4	Routine computations
5	Survey	6	Ideal gases only
7	Ideal gases and condensed phases	8	Non-ideal systems
	9	Variable temperature, pressure	

TABLE 5.5.2 *Computational results—an example*

Example: $C_3H_8 + \mathscr{R}(O_2 + 4N_2) \rightarrow$ products at 2200 °K, 40 atm.

The numbers in the table are in mole fractions: $x_i = n_i/n$, where $n = \sum_g n_g$.

Accuracy is $\pm 0.002\%$.

$\mathscr{R}$ Species	1·0	1·5	2·0	3·0	4·0	5·0	20·0
CO_2	0·00002	0·00002	0·00989	0·03404	0·06891	0·10795	0·02933
N_2	0·39996	0·46166	0·53322	0·63147	0·69554	0·73874	0·77854
H_2O	0·00018	0·00048	0·05675	0·12376	0·14830	0·14673	0·03861
CO	0·19976	0·23030	0·19006	0·12383	0·06151	0·00294	0·00007
H_2	0·39953	0·30704	0·20966	0·08657	0·02546	0·00077	0·00002
H	0·00056	0·00049	0·00041	0·00026	0·00014	0·00002	0·00000
OH	0·00000	0·00000	0·00002	0·00005	0·00012	0·00068	0·00115
O	0·00000	0·00000	0·00000	0·00000	0·00000	0·00001	0·00016
NO	0·00000	0·00000	0·00000	0·00001	0·00003	0·00097	0·01090
O_2	0·00000	0·00000	0·00000	0·00000	0·00000	0·00119	0·14182
NO_2	0·00000	0·00000	0·00000	0·00000	0·00000	0·00000	0·00007
C (graphite)	0·10020	0·00051	0·00000	0·00000	0·00000	0·00000	0·00000

THE METHOD OF LAGRANGIAN MULTIPLIERS

It is felt by the authors that a formal proof of the method of Lagrangian multipliers is beyond the scope of this book. However, readers to whom the technique is not familiar may find a verification of the method in a simple case of some interest.

Let it be desired to find the extremum of the function

$$y = x_1^2 + x_2^2$$

subject to the side condition

$$x_1 + x_2 = 1.$$

This can be solved, of course, by normal means. The side condition can be used to find x_2 as a function of x_1, i.e.

$$x_2 = 1 - x_1.$$

This can then be substituted in the expression for y to yield

$$y = x_1^2 + (1 - x_1)^2.$$

Differentiating this expression with respect to x_1, and equating the derivative to zero to find the extremum leads at once to a value of x_1 of $\frac{1}{2}$. The side condition yields a corresponding value for x_2 of $\frac{1}{2}$. Thus, conventional means lead to the solution that an extremum of y subject to the given side condition lies at $(\frac{1}{2}, \frac{1}{2})$, and the value of y (a minimum in this case) is $\frac{1}{2}$.

To use the method of Lagrangian multipliers, form first the expression

$$L = (x_1^2 + x_2^2) + \chi \, (x_1 + x_2 - 1)$$

where L is called the Lagrangian, or Lagrangian function, and where χ is the Lagrangian multiplier. Differentiating L with respect to x_1 and x_2 leads to

$$\frac{\partial L}{\partial x_1} = 2x_1 + \chi, \qquad \frac{\partial L}{\partial x_2} = 2x_2 + \chi.$$

Equating these partial derivatives simultaneously to zero leads to

$$x_1 = -\chi/2, \qquad x_2 = -\chi/2$$

(which implies that the extremal values of x_1 and x_2 are equal). These equations, connecting the extremal x_1 and x_2 values, together with the original side condition can then be used to solve first for χ, and then for x_1 and x_2. Substituting for x_1 and x_2 into the side condition leads to $\chi = -1$ which, when inserted in the extremal condition, yields $(\frac{1}{2}, \frac{1}{2})$ as the values of x_1, x_2 at which the extremum lies.

Thus the use of the Lagrangian multiplier method has led to the same result as that derived by more conventional methods.

NOMENCLATURE

a	coefficient in C_p as a function of T
a, a_i	van der Waals constants
a_e	number of gram atoms of element e per equivalent formula of reactant
a_{ie}, a_i^e	number of atoms of element e in species i
a_{sg}	$= \sum_e \bar{a}_{es} a_{ge}$ (equation 2.4.6)
A	free energy (Helmholtz free energy)
A	number of formula weights of reactant
A	first virial coefficient
$\mathscr{A}_i$	chemical formula of species i
b	coefficient in C_p as a function of T
b, b_i	van der Waals constants
b_j	non-ideality factor (equation 4.2.35)
b_{gh}	a_{ie}, corrected for presence of condensed species
B	matrix of coefficients $\sum_g b_{gh} b_{gh'} m_g$
B, B'	second virial coefficients
B_e	elemental abundance of element e
B'_e	B_e corresponding with estimated composition
B_{il}	second virial coefficient of interaction between species i and l
B_s	$= \sum_e \bar{a}_{es} B_e$ (equation 2.4.6)
$\mathscr{B}_e$	chemical symbol for element e
c	coefficient in C_p as a function of T
c_i	abbreviation for $\mu_i^\circ/RT + \ln p$ (equation 3.2.12)
C_u^r	expansion coefficients (equation 5.2.14)
C	number of components
C, C'	third virial coefficients
C_K	function for constraint K (equation 5.1.8)
$(C_p)_i, (C_v)_i$	specific heat at constant p or V, of species i
C_p	heat capacity: $C_p = \sum_i n_i (C_p)_i$ (equation 4.3.23)
d	total differential of property whose symbol follows d
d_c	trial-and-error approximation for n_c (equation 4.4.2)
D, D'	fourth virial coefficients
D_i	linear function in m_i (equation 3.3.21)
D_j	discrepancy in $(K_x)_j$ (equation 4.5.4)
D_{il}	$= (\partial \Phi_i / \partial y_l)$
E	energy
E_i	linear function in m_i (equation 3.3.22)
f_i, f	partial, total fugacity
F	variance, number of degrees of freedom

F_k — residuals or discrepancies, due to estimates, for equation k

g — gravitational constant

Δg_e — $= \sum_m \nu_{me} n_m \, \Delta g_m$ (equation 4.3.22)

Δg_j — $= (\Delta \, G/RT)_j$

G — Gibbs free energy

G_k — functions of mole fractions (equation 5.1.3)

GDF — Gibbs free energy deviation function (equation 1.5.4)

H — enthalpy

I — inverse matrix of B

I_s — specific impulse

k — proportionality constant

$(K_p)_r$ — equilibrium constant, in partial pressures, for reaction r

$(K_x)_r$ — equilibrium constant, in mole fractions, for reaction r

K_j — abbreviation involving $(K_p)_j$ and p (equation 4.2.21)

L — Lagrangian function

L_c — order of polynomial in y_c (equation 4.3.28)

m_i, m — estimates of n_i, n

M — number of elements

n — sum of number of gaseous moles $n = \sum_g n_g$

n_i — number of moles (or gram atoms) of species i

n_R — number of ratios of elemental abundances to be specified (equation 1.3.18)

N — total number of species

p — total pressure

p_i — partial pressure of species i

P — number of points on x, β curve (equation 3.2.16)

q_c — linear function of the a_{ie} and B_e (equation 2.3.6)

q_e — $= \sum_m \nu_{me} n_m \, q_m$ (equation 4.3.21)

q_j — $= \left(\dfrac{\partial \ln K_p}{\partial \ln T} \right)_j$

$Q(n_i)$ — quadratic approximation to $G(n_i)$

Q_h — linear combination of the B_e

Q_j — estimate of $(K_p)_j$ (equation 4.3.3) or of $(K_x)_j$ (equation 4.5.4)

Q_p, Q_v — heat of reaction at constant pressure, volume

Q_{sys}, Q_{sur} — heat contents of system, surroundings (equation 1.2.17)

r — reaction index

r — iteration step number

r_{ck} — abbreviation $\sum_i a_{ik} \nu_{ic} y_i$ (equation 4.2.26)

r_{ef} — abbreviation $\sum_m \nu_{me} \nu_{mf} n_m$ (equation 4.3.20)

r_k — positive values in a sequence k (equation 3.2.9)

R — gas constant

R — number of independent reactions that can take place (equation 1.3.14)

R' — number of unrestricted independent reactions (equation 1.3.13)

R_e	$= B_e/B_r$ or $B_e/\sum\limits_e B_e$
$\mathscr{R}$	molar ratio air/propane
s_c	a linear function of R_e and a_{ie} (equation 4.2.16)
s_i	partial entropy of species i
S	entropy
S	number of pure condensed species
T	absolute temperature
T_0	initial (reactant) temperature
T_q	equilibrium temperature
v_e	rocket exhaust velocity
V	total volume
V_i	partial volume of species i
w_i, w_j	weighting factors
W	function of the w_i
x	$\sum\limits_c x_c + \sum\limits_j x_j$
x_i	mole fraction of species i
y	$\sum\limits_c y_c + \sum\limits_j y_j$
y_i	estimate of x_i
Z	number of constraints on atomic ratios (equation 1.3.13)

SUPERSCRIPTS

E	thermodynamic excess function (equation 1.4.20)
m	mixed state (equation 1.2.9)
o	standard state (or pure species)
r	iteration step number
u	unmixed state (equation 1.2.8)
ϕ	phase $(1, \ldots, \Phi)$
prime	denotes: values of parameter in initial system or value corresponding with an initial estimate

SUBSCRIPTS

c	component $(1, \ldots, C)$
e, f	element $(1, \ldots, M)$
f	for formation reaction
g	gaseous species $(1, \ldots, N-S)$
h, h'	gaseous atom balances $(1, \ldots, M-S)$
i, l	species $(1, \ldots, N)$
j	derived species $(1, \ldots, N-M)$
k	discrepancy equation index
K	constraint index $(1, \ldots, N+1)$ (equation 5.1.8)
m	gaseous molecular compound $(1, \ldots, N-M-S)$
p	denoting fitted points (LP) $(1, \ldots, P)$
q	equilibrium
r	reaction
r	reference element
s	condensed species $(1, \ldots, S)$
s	denotes starter value of subscripted mole fraction (equation 5.1.3)

T	at temperature T
u, v	auxiliary index for i $(1, \ldots, N)$ (equations 5.2.10–5.2.12)
298	at 298·15 °K ($= 25$ °C)

GREEK LETTERS

α	covolume function
α_i	formula vector of species i
β_i	$= x_i \ln x_i$
γ_i	activity coefficient of species i
Γ_i	abbreviation for $n_i \left(c_i + \ln \dfrac{n_i}{n} \right)$ (equation 3.3.32)
∂	partial differential
δ	first-order error or change
δ_{il}, δ_i^l	Kronecker delta (0 for $i \neq l$; 1 for $i = l$)
Δ	finite difference or discrepancy in property whose symbol follows Δ
ϵ, ϵ_r	extent of reaction (r)
ϵ_{il}	$= \sum\limits_j \nu_{ij}\, \nu_{lj}$ (equation 3.4.12)
ζ_i	denotes mixed or unmixed: $\zeta_i = 1$ for species i in mixed phase; $\zeta_i = 0$ for a pure species i
η_e	linear element of Lagrangian multiplier χ_e (equation 3.3.17)
θ	positive integral index (equation 4.4.3)
θ	abbreviation for $298\cdot15/T$
λ	step size, or search parameter, or scale parameter
λ_u	eigenvalue (equation 5.2.13)
μ_e	element potential for element e
μ_i	chemical potential of species i
μ_i°	standard chemical potential of species i
ν_i, ν_l	reaction coefficients for a single reaction involving species i and l
ν_{ic}, ν_i^c	stoichiometric coefficients
ξ_i	$= \ln n_i$
ξ_j	$= \sum\limits_c \nu_{jc} - 1$
π_{iu}	function defined by equation 5.2.11
$\prod\limits_c$	denotes: product over c
ρ_{lu}	function defined by equation 5.2.10
σ	step size parameter (equation 3.3.14)
$\sum\limits_i$	denotes: sum over index i
τ	tolerance, or required accuracy
ϕ	phase number
Φ	number of phases
$\Phi(n_i)$	function of the n_i
χ_c	Lagrangian multiplier for constraint characterized by subscript c
ψ	$= G/RT$
ψ_i	$= \mu_i^\circ/RT$
ω_e	linear element of Lagrangian multiplier χ_e (equation 3.3.17)

BIBLIOGRAPHY

Many of the references below are to articles appearing in the Proceedings of the First and Second Conferences on the 'Kinetics, Equilibria and Performance of High Temperature Systems'. Of these, the first one was published by Butterworths (1960) and the second one by Gordon and Breach (1963). Proceedings of a Third Conference have been published very recently by Gordon and Breach (1968). U.S. Government Reports (PB and AD) are available from the Clearinghouse for Federal Scientific and Technical Information, Springfield, Va.

Alcock, C. B. (1966). Free Energy Functions and Deviations in Thermochemical Data Storage for Inorganic Compounds. *Nature*, **209**, 198–9.

American Institute of Chemical Engineers (1965). Program for the Computer Estimation of Physical Properties. *Am. Inst. Chem. Engnrs.* New York.

Anthony, R. G. and Himmelblau, D. M. (1963). Calculation of Complex Chemical Equilibria by Search Techniques. *J. Phys. Chem.* **67**, 1080–3.

Aris, R. (1963). The Fundamental Arbitrariness in Stoichiometry. *Chem. Eng. Sci.* **18**, 554–5.

Aris, R. (1965). Prolegomena to the Rational Analysis of Systems of Chemical Reactions. *Arch. Rational Mech. and Anal.* **19**, 81–99.

Bahn, G. S. (1960). Hand Calculation of Equilibrium Compositions as a Learned Habit, and Speed-up Effected with the IBM-610 Computer. *Proc. 1st Conf. High Temp. Systems*, 137–40.

Baibuz, V. F. (1962). Calculation of the Equilibrium Composition of Gas Mixtures at High Temperatures. *Russ. J. Phys. Chem.* **36**, 751–4.

Barnhard, P. and Hawkins, A. W. (1963). Singularities Occurring in the Newton–Raphson Solution of Chemical Equilibria. *Proc. 2nd Conf. High Temp. Systems*, 235–41.

Battelle Memorial Institute (1949). Physical Properties and Thermodynamic Functions of Fuels, Oxidizers and Products of Combustion. *Project RAND Repts.*
  I   Fuels, AD-605 967.
  II Oxidizers, AD-605 966.

Beattie, J. A. (1949). The Computation of the Thermodynamic Properties of Real Gases and Mixtures of Real Gases. *Chem. Rev.* **44**, 141–92.

Boas, A. H. (1963). Modern Mathematical Tools for Optimization. *Chem. Engng. Reprint.* (Available from Chemical Engineering, McGraw Hill Publications, N.Y.)

Bockris, J. O'M., White, J. L. and Mackenzie, J. D. (1959). *Physicochemical Measurements at High Temperatures.* Butterworths Scient. Publ., London.

Boll, R. H. (1961). Calculation of Complex Equilibrium With an Unknown Number of Phases. *J. Chem. Phys.* **34**, 1108–10.

Boynton, F. P. (1960). Chemical Equilibrium in Multi-Component Poly-phase Systems. *J. Chem. Phys.* **32**, 1880–1.
Boynton, F. P. (1963). Computation of Equilibrium Composition and Properties in a Gas Obeying the Virial Equation of State. *Proc. 2nd Conf. High Temp. Systems*, 187–204.
Brandmeier, H. E. and Harnett, J. J. (1960). Performance Calculations for Reaction Engines. II A Brief Survey of Past and Current Methods of Solution for Equilibrium Composition. *Proc. 1st Conf. High Temp. Systems*, 69–73.
Brinkley, S. R. (1946). Note on the Conditions of Equilibrium for Systems of Many Constituents. *J. Chem. Phys.* **14**, 563–4, 686.
Brinkley, S. R. (1947). Calculation of the Equilibrium Composition of Systems of Many Constituents. *J. Chem. Phys.* **15**, 107–10.
Brinkley, S. R. (1956). Computational Methods in Combustion Calcu-lations. From B. Lewis *et al. High Speed Aerodynamics and Jet Propulsion—Combustion Processes*, Vol. II, 64–98. Princeton Univ. Press.
Brinkley, S. R. (1960). Calculation of the Thermodynamic Properties of Multicomponent Systems and Evaluation of Propellant Performance Parameters. *Proc. 1st Conf. High Temp. Systems*, 74–81.
Brinkley, S. R. (1966). *On the Calculation of the Equilibrium Composition and Thermodynamic Properties of Multiconstituent Systems*, paper presented at 16th Canadian Chem. Eng. Conf., Windsor, Ontario.
Browne, H. N., Williams, M. M. and Cruise, D. R. (1960). The Theo-retical Computation of Equilibrium Compositions, Thermodynamic Properties and Performance Characteristics of Propellant Systems. *NAVWEPS Rept.* 7043, NOTS-Tech. Publ. 2434.
Calvin, M. (1965). Chemical Evolution. *Proc. Roy. Soc.* A **288**, 441–66.
Chu, S. T. (1958). An Iterative Method of Determining Equilibrium Compositions of Reacting Gases. *Jet Prop.* **28**, 252–4.
Clasen, R. J. (1965). *The Numerical Solution of the Chemical Equilibrium Problem*. Rand Corp. Memo 4345-PR, AD-609 904.
Cook, M. A. (1958). *The Science of High Explosives*. Reinhold Publ. Corp. New York.
Core, T. C., Saunders, S. G. and McKittrick, P. S. (1963). Versatile Specific Impulse Program for the IBM-650 and -704 Computers. *Proc. 2nd Conf. High Temp. Systems*, 243–60.
Courant, R. (1957). *Differential and Integral Calculus*. Vol. II. Inter-science, New York.
Crout, P. D. (1941). A Short Method for Evaluating Determinants and Solving Systems of Linear Equations with Real or Complex Co-efficients. *AIEE Trans. (Suppl.)* **60**, 1235–41.
Cruise, D. R. (1964). Notes on the Rapid Computation of Chemical Equilibria. *J. Phys. Chem.* **68**, 3797–802.
Damköhler, G. and Edse, R. (1943). The Composition of Dissociating Combustion Gases and the Calculation of Simultaneous Equilibria. *Z. Elektrochem.* **49**, 178–86.
Dantzig, G. B. and De Haven, J. C. (1962). On the Reduction of Certain Multiplicative Chemical Equilibrium Systems to Mathematically Equivalent Additive Systems. *J. Chem. Phys.* **36**, 2620–7.

Davidon, W. C. (1959). Variable Metric Method for Minimization. *A.E.C. Res. and Dev. Rept.* ANL-5990 (Rev.) Argonne Nat'l. Lab., Argonne (Ill.), U.S.A.

Desré, P., Durand, F. and Bonnier, E. (1964). Matrix Presentation of Thermodynamic Functions of a Reaction for Automatic Equilibrium Calculation. I: Heterogeneous Equilibria Between a Gaseous Phase and Pure Condensed Constituents. *Rev. Hautes Tempér. et Réfract.* 1, 321–4.

Donegan, A. J. and Farber, M. (1956). Solution of Thermochemical Propellant Calculations on a High-Speed Digital Computer. *Jet. Prop.* 26, 164–71.

Eck, R. V., Lippincott, E. R., Dayhoff, M. O. and Pratt, Y. T. (1966). Thermodynamic Equilibrium and the Inorganic Origin of Organic Compounds. *Science*, 153, 628–33.

Fiacco, A. V. and McCormick, G. P. (1963). Programming Under Nonlinear Constraints by Unconstrained Minimization—A Primal–Dual Method. *Research Analysis Corp. Report TP-96.* McLean, Va.

Fiacco, A. V. and McCormick, G. P. (1964a). The SUMT for Nonlinear Programming—A Primal–Dual Method. *Mgt. Sci.* 10, 360–6.

Fiacco, A. V. and McCormick, G. P. (1964b). Computational Algorithm for the SUMT for Nonlinear Programming. *Mgt. Sci.* 10, 601–17.

Fiacco, A. V. and McCormick, G. P. (1966). Extensions of SUMT for Nonlinear Programming: Equality Constraints and Extrapolation. *Mgt. Sci.* 12, 816–28.

Fiacco, A. V. and McCormick, G. P. (1967). The SUMT Without Parameters. *Oper. Res.* 15, 820–7.

Finerman, A. and Revens, L. (1964, 1966, 1967). Permuted and Subject Index to 'Computing Reviews'. *Ass. Comp. Mach.*

Fletcher, R. (1965). Function Minimization Without Evaluating Derivatives—A Review. *Comp. J.* 8, 33–41.

Fletcher, R. and Powell, M. J. D. (1963). A Rapidly Convergent Descent Method for Minimization. *Comp. J.* 6, 163–8.

Freudenstein, F. and Roth, B. (1963). Numerical Solution of Systems of Nonlinear Equations. *J. Ass. Comp. Mach.* 10, 550–6.

Gibbs, J. W. (1961). *The Scientific Papers.* Dover Publ. Inc., New York.

Glass, H. and Cooper, L. (1965). Sequential Search: A Method for Solving Constrained Optimization Problems. *J. Ass. Comp. Mach.* 12, 71–82.

Glasstone, S. and Lewis, D. (1960). *Elements of Physical Chemistry.* D. van Nostrand Co. Inc., p. 328.

Goldfarb, D. and Lapidus, L. (1967). *Conjugate Gradient Method for Nonlinear Programming.* Paper presented at 61st Nat'l. Meeting of Am. Inst. Chem. Engrs., Houston, Texas. (*Optimization—Recent Advances in Theory and Application*, Part II, unpublished.)

Goldwasser, S. R. (1959). Basis for Calculating Equilibrium Gas Composition on a Digital Computer. *Ind. Eng. Chem.* 51, 595–6.

Gordon, S., Zeleznik, F. J. and Huff, V. N. (1959). A General Method for Automatic Computation of Equilibrium Compositions and Theoretical Rocket Performance of Propellants. *NASA Tech. Note* D-132.

Guggenheim, E. A. (1952). *Mixtures.* Clarendon Press, Oxford.

166   COMPUTATION OF CHEMICAL EQUILIBRIA

Hancock, H. J. and Motzkin, T. S. (1960). Analysis of the Mathematical Model for Chemical Equilibrium. *Proc. 1st Conf. High Temp. Systems,* 82–9.

Harker, H. J. (1967). The Calculation of Equilibrium Flame Gas Compositions. *J. Inst. Fuel,* **40,** 206–13.

Higman, B. (1955). *Applied Group Theoretic and Matrix Methods.* Oxford Univ. Press.

Hildebrand, F. B. (1956). *Introduction to Numerical Analysis.* McGraw Hill Book Co., New York.

Hilsenrath, J., Klein, M. and Sumida, D. Y. (1959). The Calculation of the Equilibrium Composition and Thermodynamic Properties of Dissociated and Ionized Gaseous Systems. *Symp. Thermodyn. Transport Properties of Gases, Liquids, Solids.* McGraw Hill Book Co., New York, 416–37.

Hooke, R. and Jeeves, T. A. (1961). Direct Search Solution of Numerical and Statistical Problems. *J. Ass. Comp. Mach.* **8,** 212–29.

Horn, F. and Schüller, W. (1957). On the Calculation of the Composition and Thermodynamic Functions of Dissociating Combustion Gases. *Dechema Monogr.* **29,** 143–64.

Horn, F. and Troltenier, U. (1962). Calculation of Simultaneous Chemical Equilibrium with a Programmed Computer. *Chem. Ingr.-Techn.* **34,** 551–6.

Huff, V. N., Gordon, S. and Morrell, V. E. (1951). General Method and Thermodynamic Tables for Computation of Equilibrium Composition and Temperature of Chemical Reactions. *Nat'l. Adv. Committee Aeronautics (NACA) Report* 1037.

Janz, G. J. (1958). *Estimation of Thermodynamic Properties of Organic Compounds.* Academic Press, New York.

Jones, A. P. (1967). The Chemical Equilibrium Problem: An Application of SUMT. *Research Analysis Corp. Tech. Paper* 272, AD-819 848.

Kaeppeler, H. J. and Baumann, G. (1957). Systems with Chemically Reacting Components in Equilibrium. I. Calculation of the Composition of the Mixture. *Astronaut. Acta (Vienna)* **3,** 28–46.

Kandiner, H. J. and Brinkley, S. R. (1950). Calculation of Complex Equilibrium Relations. *Ind. Eng. Chem.* **42,** 850–5.

Kelley, K. K. (1960). High Temperature Heat-Content, Heat-Capacity, and Entropy Data for the Elements and Inorganic Compounds. *U.S. Bur. Mines Bull. No.* 584.

Kelley, K. K. (1961). Entropies of the Elements and Inorganic Compounds. *U.S. Bur. Mines Bull. No.* 592.

Kelley, K. K. (1962). A Reprint of Bulletins 383, 384, 393, and 406. *U.S. Bur. Mines Bull. No.* 601.

Kobe, K. A. and Leland, T. W. (1954). The Calculation of Chemical Equilibrium in a Complex System. *Univ. of Texas Bur. Eng. Res. Special Publ. No.* 26. Univ. Texas, Austin, Texas.

Koopmans, T. C. (1951). *Activity Analysis of Production and Allocation.* J. Wiley & Sons, New York.

Krieger, F. J. and White, W. B. (1948). A Simplified Method for Computing the Equilibrium Composition of Gaseous Systems. *J. Chem. Phys.* **16,** 358–60.

BIBLIOGRAPHY 167

Kubaschewski, O. and Evans, E. Ll. (1958). *Metallurgical Thermo-Chemistry*. 3rd ed. Pergamon Press, London.

Kubert, B. R. and Stephanou, S. E. (1960). Extension to Multiphase Systems of the RAND Method for Determining Equilibrium Compositions. *Proc. 1st Conf. High Temp. Systems*, 166–70.

Kunz, K. S. (1957). *Numerical Analysis*. McGraw Hill Book Co., New York.

Lanczos, C. (1956). *Applied Analysis*. Prentice Hall Inc., Englewood Cliffs, N.J.

Landolt-Börnstein (1961). *Zahlenwerte und Funktionen*, II. Part 4: *Kalorische Zustandsgrössen*. Springer-Verlag, Berlin.

Lavi, A. and Vogl, T. P. (1966), editors, *Recent Advances in Optimization Theory*. J. Wiley & Sons, New York.

Leon, A. (1965). An Annotated Bibliography on Optimization. *Mental Health Res. Inst.*, Preprint No. 162. Univ. Michigan, Ann Arbor, Mich., U.S.A. See also Lavi & Vogl (1966), pp. 599–649.

Levine, H. B. (1962). Chemical Equilibrium in Complex Mixtures. *J. Chem. Phys.* 36, 3049–50.

Lewis, G. N. and Randall, M. (1961). *Thermodynamics*. 2nd ed. McGraw Hill Book Co., New York.

Lu, C. S. (1967). A Thermodynamic Study of Multiple Reaction Systems at and Near Equilibrium. *Diss. Cal. Inst. Tech.* 67–6160, *Univ. Microfilms Inc.* Ann Arbor, Michigan.

Marek, J. and Holub, R. (1964). The Calculation of Complex Chemical Equilibria with a Digital Computer. *Coll. Czech. Chem. Commun.* 29, 1085–93.

Martin, F. J. and Yachter, M. (1951). Calculation of Equilibrium Gas Compositions. *Ind. Eng. Chem.* 43, 2446–9.

McBride, B. J. and Gordon, S. (1967). FORTRAN IV Program for the Calculation of Thermodynamic Data. *NASA*, TN D-4097.

McBride, B. J., Heimel, S., Ehlers, J. G. and Gordon, S. (1963). Thermodynamic Properties to 6000 °K for 210 Substances, Involving the First 18 Elements. *NASA*, SP-3001. U.S. Dept. Commerce, Washington, D.C.

McEwan, W. S. (1950). Equilibrium Composition and Thermodynamic Properties of Combustion Gases. *NOTS Rept.* 289, PB-129 595.

McEwan, W. S. and Skolnik, S. (1951). An Analog Computer for Flame Gas Composition. *Rev. Sci. Instr.* 22, 125–32.

McGee, H. A. and Heller, G. (1962). Plasma Thermodynamics I: Properties of Hydrogen, Helium and Lithium as Pure Elemental Plasmas. *J. Amer. Rocket Soc.* 32, 203–15.

McMahon, D. G. and Roback, R. (1960). Machine Computation of Chemical Equilibria in Reacting Systems. *Proc. 1st Conf. High Temp. Systems*, 105–14.

Mentz, R. M. (1960). Programme for Computing Equilibrium Temperature and Composition of Chemical Reactions. *Proc. 1st Conf. High Temp. Systems*, 115–22.

Michels, H. H. and Schneiderman, S. B. (1963). Chemical Equilibria in Real Gas Systems. *Proc. 2nd Conf. High Temp. Systems*, 205–34.

Mingle, J. O. (1962). A Technique for Computer Solution of Complex Chemical Equilibrium Problems. *Kansas State Univ. Eng. Expt. Sta. Spec. Rept. No.* 25.

Naphtali, L. M. (1959). Complex Chemical Equilibria by Minimizing Free Energy. *J. Chem. Phys.* **31**, 263–4.
Naphtali, L. M. (1960). Computing Complex Chemical Equilibria by Minimizing Free Energy. *Proc. 1st Conf. High Temp. Systems*, 181–3.
Naphtali, L. M. (1961). Calculate Complex Chemical Equilibria with a Technique Based on Minimizing Free Energy. *Ind. Eng. Chem.* **53**, 387–8.
Neumann, K. K. (1962). Discussion of Error in the Calculation of Simultaneous Equilibria. *Progr. Int. Research on Thermodyn. and Transport Properties*, 209–17. Academic Press, New York.
Neumann, K. K. (1966). Calculation of Simultaneous Equilibria by an Iteration Method in Matrix Notation. *Brennstoff-Chemie* **47**, 146–9.
Nikolskii, S. S. (1966). On the Thermodynamics of Homogeneous Mixtures with Transformable Components. *Theor. Exp. Chem. (Ukr. SSR)* **2**, 343–52.
Oliver, R. C., Stephanou, S. E. and Baier, R. W. (1962). Calculating Free Energy Minimization. *Chem. Engng.* **69**, (4), 121–8.
Passy, U. and Wilde, D. J. (1968). A Geometric Programming Algorithm for Solving Chemical Equilibrium Problems, S.I.A.M. *J. Appl. Math.* **16**, 363–73.
Peneloux, A. (1949). Remarks on the Definition of a Chemical System with Stoichiometric Equations. *Compt. Rend. Ac. Sci. (Paris)*, **228**, 1727–9.
Piehler, J. (1962). Calculation of Chemical Equilibria, *Chem. Tech. (Berlin)*, **14**, 408–10.
Pings, C. J. (1961). Thermodynamics of Chemical Equilibrium. I: Effect of Temperature and Pressure. *Chem. Eng. Sci.* **16**, 181–8.
Pings, C. J. (1963). Thermodynamics of Chemical Equilibrium. II: Effect of Volume, Entropy and Enthalpy. *Chem. Eng. Sci.* **18**, 671–6.
Pings, C. J. (1966). Thermodynamics of Chemical Equilibrium. III: Effect of Heat, Work and Viscous Dissipation. *Chem. Eng. Sci.* **21**, 693–5.
Pings, C. J. (1964). Optimization of Initial Composition in Adiabatic Equilibrium Gas-Phase Reactions. *Am. Inst. Chem. Engrs. J.* **10**, 934–6.
Potter, R. L. and Vanderkulk, W. (1960). Computation of the Equilibrium Composition of Burnt Gases. *J. Chem. Phys.* **32**, 1304–7.
Powell, H. N. and Sarner, S. F. (1959). The Use of Element Potentials in Analysis of Chemical Equilibrium. Vol. 1. *General Electric Co.*, Report R59/FPD 796.
Prigogine, I. and Defay, R. (1947). On the Number of Independent Constituents and the Phase Rule. *J. Chem. Phys.* **15**, 614–5.
Prigogine, I. and Defay, R. (1954). *Chemical Thermodynamics*. Longmans, Green & Co., London.
Rosen, J. B. (1960). The Gradient Projection Method for Non-linear Programming. Part 1: Linear Constraints. *J. Soc. Ind. Appl. Math.* **8**, 181–217.
Rossini, F. D., Wagman, D. D., Evans, W. H., Levine, S. and Jaffe, I. (1952). Selected Values of Chemical Thermodynamic Properties. *Nat'l. Bur. Standards Circular* **500**. U.S. Gov't. Print Office, Washington, D.C.

Sachsel, G. F., Mantis, M. E. and Bell, J. C. (1949). A Note on The Calculation of Multicomponent Propellant Gas Compositions, *3rd Symp. Combustion*, 620–3. Williams & Wilkins Co., Baltimore, Md.

Scarborough, J. B. (1930). *Numerical Mathematical Analysis*. Johns Hopkins Press, Baltimore, Md.

Schott, G. L. (1964). Computation of Restricted Equilibria by General Methods. *J. Chem. Phys.* **40,** 2065–6.

Scully, D. B. (1962). Calculation of the Equilibrium Compositions for Multiconstituent Systems. *Chem. Eng. Sci.* **17,** 977–85.

Shapiro, N. Z. (1964*a*). On the Behaviour of a Chemical Equilibrium System When its Free Energy Parameters are Changed. *RAND Corp. Memo.* 4128-PR, AD-600 884.

Shapiro, N. Z. (1964*b*). A Generalized Technique for Eliminating Species in Complex Chemical Equilibrium Calculations. *RAND Corp. Memo* 4205-PR, AD-605 316.

Shapiro, N. Z. and Shapley, L. S. (1964). Mass Action Laws and the Gibbs Free Energy Function. *RAND Corp. Memo* 3935-1-PR, AD-605 919.

Shear, D. B. (1968). Stability and Uniqueness of the Equilibrium Point in Chemical Reaction Systems, *J. Chem. Phys.* **48,** 4144–7.

Shelton, R. A. J. and Blairs, S. (1966). Thermochemical Data Storage for Inorganic Compounds. *Nature*, **211,** 1397–8.

Skinner, H. A. (1964). Key Heat of Formation Data. *Pure. Appl. Chem.* **8,** 113–30.

Smith, W. R. and Missen, R. W. (1968). Calculating Complex Chemical Equilibria by an Improved Reaction-Adjustment Method. *Can. J. Chem. Eng.* **46,** 269–72.

Snow, R. H. (1963). Computer Procedures for Determining Chemical Equilibrium. *I.I.T. Res. Inst. Rept.* C929-3. Chicago, Ill.

Spang, H. A. (1962). A Review of Minimization Techniques for Non-linear Functions. *S.I.A.M. Rev.* **4,** 343–65.

Stone, E. E. (1966). Complex Chemical Equilibria. *J. Chem. Educ.* **43,** 241–4.

Storey, S. H. (1965). The Effect of Changes in Initial Composition on the Equilibria of Solid-Gas Systems. *Can. J. Chem. Eng.* **43,** 168–72.

Storey, S. H. and van Zeggeren, F. (1964). Computation of Chemical Equilibrium Compositions. *Can. J. Chem. Eng.* **42,** 54–5.

Storey, S. H. and van Zeggeren, F. (1967). Solving Complex Chemical Equilibria by a Method of Nested Iterations. *Can. J. Chem. Eng.* **45,** 323–6.

Storey, S. H. and van Zeggeren, F. (1969). Computation of Chemical Equilibrium Compositions. II, *Can. J. Chem. Eng.* (to be published).

Stull, D. R. (1965). *JANAF Thermochemical Tables*, PB 168 370 + supplements I (1966), II (1967) and III (1968).

Sutton, G. P. (1963). *Rocket Propulsion Elements.* 3rd edition. J. Wiley & Sons, New York.

Traub, J. F. (1964). *Iterative Methods for the Solution of Equations.* Prentice Hall Inc., Englewood Cliffs, N.J.

Tsao, C. C. and Wiederhold, G. (1963). A Computer Program for the

170   COMPUTATION OF CHEMICAL EQUILIBRIA

Evaluation of Rocket Fuels. *Proc. 2nd Conf. High Temp. Systems*, 261–9.

van Zeggeren, F. and Storey, S. H. (1969). The Effect of Changes in Initial Reactant Composition on Solid–Gas Equilibria Resulting from Constant-Volume, Adiabatic Processes. *Can. J. Chem. Eng.* **47**, 81–4.

Villars, D. S. (1959). A Method of Successive Approximations for Computing Combustion Equilibria on a High Speed Digital Computer. *J. Phys. Chem.* **63**, 521–5.

Villars, D. S. (1960). Computation of Complicated Combustion Equilibria on a High-Speed Digital Computer. *Proc. 1st Conf. High Temp. Systems*, 141–51.

Von Stein, M. (1943). Method of Calculating Flame Temperatures, Enthalpy and Entropy of Combusion Gases. *Forsch. Gebiete Ingenieurw.* **14**, 113–23.

Von Stein, M. and Voetter, H. (1953). The Calculation of Simultaneous Equilibria. *Z. Elektrochem.* **57**, 119–24.

Wagman, D. D., Evans, W. H., Parker, V. B., Halow, I., Baily, S. M. and Schumm, R. H. (1968). Selected Values of Chemical Thermodynamic Properties. *Nat'l. Bur. Standards Tech. Note* 270–3.

Warga, J. (1963). A Convergent Procedure for Solving the Thermochemical Equilibrium Problem. *J. Soc. Indust. Appl. Math.* **11**, 594–606.

Weinberg, F. J. (1957). Explicit Equations for the Calculation, by Successive Approximations, of Equilibrium Gas Compositions at High Temperatures. *Proc. Roy Soc.* A **241**, 132–40.

White, W. B. (1967). Numerical Determination of Chemical Equilibrium and the Partitioning of Free Energy. *J. Chem. Phys.* **46**, 4171–5.

White, W. B., Johnson, S. M. and Dantzig, G. B. (1958). Chemical Equilibrium in Complex Mixtures. *J. Chem. Phys.* **28**, 751–5.

Wiederkehr, R. R. V. (1962). Matrix Representation of Thermodynamic Properties of Multicomponent Systems. *J. Chem. Phys.* **37**, 1192–9.

Wilde, D. J. and Beightler, C. S. (1967). *Foundations of Optimization.* Prentice-Hall Inc., Englewood Cliffs, N.J.

Wilkes, M. V. (1956). *Automatic Digital Computers.* J. Wiley & Sons, New York.

Wilkins, R. L. (1960). Note on the Linearization Method for Computing Chemical Equilibrium in Complex Systems. *Proc. 1st Conf. High Temp. Systems*, 123–7.

Winternitz, P. F. (1949). A Method for Calculating Simultaneous, Homogeneous Gas Equilibria and Flame Temperatures. *3rd Symp. Combustion*, 623–33. Williams & Wilkins Co., Baltimore, Md.

Zeleznik, F. J. and Gordon, S. (1960). An Analytical Investigation of Three General Methods of Calculating Chemical Equilibrium Compositions. *NASA*, Techn. Note D-473.

Zeleznik, F. J. and Gordon, S. (1961). Simultaneous Least-Squares Approximation of a Function and its First Integrals with Application to Thermodynamic Data. *NASA*, Techn. Note D-767.

Zeleznik, F. J. and Gordon, S. (1962). A General IBM-704 or 7090 Program for Computation of Chemical Equilibrium Compositions,

Rocket Performance, and Chapman-Jouguet Detonations. *NASA*, Techn. Note D-1454.

Zeleznik, F. J. and Gordon, S. (1968). Calculation of Complex Chemical Equilibria. *Ind. Eng. Chem.* **60,** 27–57.

Zemansky, M. W. (1957). *Heat and Thermodynamics.* McGraw Hill Book Co., New York.

AUTHOR INDEX

SUBJECT INDEX